The Ethics of the
Climate Crisis

T0165131

The Ethics of the Climate Crisis

Robin Attfield

polity

The right of Robin Attfield to be identified as Author of this Work has been asserted in accordance with the UK Copyright, Designs and Patents Act 1988.

First published in 2024 by Polity Press

Polity Press
65 Bridge Street
Cambridge CB2 1UR, UK

Polity Press
111 River Street
Hoboken, NJ 07030, USA

ISBN-13: 978-1-5095-5908-4
ISBN-13: 978-1-5095-5909-1(pb)

A catalogue record for this book is available from the British Library.

Library of Congress Control Number: 2023945830

Typeset in 11 on 13pt Sabon
by Fakenham Prepress Solutions, Fakenham, Norfolk NR21 8NL
Printed and bound in Great Britain by CPI Group (UK) Ltd, Croydon

The publisher has used its best endeavours to ensure that the URLs for external websites referred to in this book are correct and active at the time of going to press. However, the publisher has no responsibility for the websites and can make no guarantee that a site will remain live or that the content is or will remain appropriate.

Every effort has been made to trace all copyright holders, but if any have been overlooked the publisher will be pleased to include any necessary credits in any subsequent reprint or edition.

For further information on Polity, visit our website:
politybooks.com

Contents

Acknowledgements

I am grateful to Brian Jackson for looking over chapter 2, to Anna Wienhues for her detailed comments on chapter 3, and to her and other members of the European Group of the International Society for Environmental Ethics for comments on a section of chapter 5. Thanks also go to Workineh Kelbessa, who has authorized me to refer to an unpublished chapter of his. The staff of Polity Press have once again served me well, and I am again grateful for their excellent support. I am also grateful to two anonymous referees and to an anonymous proofreader for rescuing me from various pitfalls.

Above all, thanks are due to my wife Leela. She has recently published two books herself, a collection of short stories, *Fresh Beginnings* (Bridge House, 2022), and the novel *A Distant Voice in the Darkness* (186 Publishing): good reads, both. Without her continual care, I would have been unable to begin, compose or complete this book.

1

Introduction

Most people now accept that climate change is a genuine and serious threat both to our own generation and to coming ones. There is a scientific near-consensus to this effect (Houghton 2015). Yet it is far from being accepted that we are ethically obliged to take action, whether 'we' stands for citizens, for companies, for local authorities, or for national governments.

This book is written to present the ethical case for climate action: for action, that is, on the part of individuals and households, of companies and other non-governmental organizations, and of governments at all levels, from local government, via regional governments to national governments and to international organizations. It is inadequate if we applaud the policies agreed at conferences like COP26, the Glasgow international conference of 2021 (BBC 2021), but then forget and ignore what we are committed to. And it is a dereliction if citizens assume that our governments have these matters in hand, and focus on other issues to the effective exclusion of climate matters, leaving the drift towards climate catastrophe largely unaffected. (Peter Singer reaches similar conclusions: see Singer 2023: 295–6.) It is also insufficient if, as citizens, we focus only on one form of action (such as offsetting, as some ethicists do) and ignore other forms of household action and our potential role as campaigners, signers of petitions and lobbyists. It is also a dereliction if governments agree to contribute to international climate objectives but ignore them in their domestic and foreign policies.

The exact nature of the ethical case is to some extent controversial, particularly where the content and extent of climate justice are concerned. At the same time, the issues are urgent, as the phrase 'climate crisis' in the title of this book indicates. In this book, the ethical case will be examined, and related conclusions about the responsibilities of individuals, corporations and governments will be explicitly drawn. Ethicists have been applying ethical theory to practical issues for at least the last fifty years (as used to be done in the more distant past by philosophers like Plato and Spinoza). Readers who would like to see another short book on global ethics, maybe to get used to this kind of writing, may like to look at Singer's admirable book *One World* (Singer 2002).

A crisis is a time of intense difficulty or danger (in the original Greek, a time of judgement). In medical situations, it is the turning point of a disease when an important change takes place, indicating either recovery or death. To develop the medical metaphor, but to return to the climate crisis, it is a time when a whole range of natural systems are on course to reach a tipping point, defined as 'the point at which small changes become significant enough to cause a larger, more critical change that can be abrupt, irreversible and lead to cascading effects' (Cho 2021: 2; Cho is cited several times in this and the next chapter, and has, it is worth adding, authored over 200 papers, the great majority of which will have successfully undergone peer review). Under sustained stress, systems 'become increasingly likely to reach "a critical threshold beyond which a system reorganizes, often abruptly and/or irreversibly"' (IPCC 2021: 28, cited in Caldecott 2022a: 3). Several natural systems could interact, as one or more reaches a tipping point, with a cascade of unpalatable yet unpredictable climatic events ensuing.

The Arctic ice cap has been melting, as also has the ice sheet that covers Greenland. Sea levels and ocean levels are rising, threatening small islands and coastal cities and communities. The permafrost of Siberia is at risk of melting enough to release large quantities of methane, a greenhouse gas much more potent than carbon dioxide. The West Antarctic ice sheet has lost trillions of tonnes of ice already this century, and is in danger of reducing or even disappearing. The circulation system of Atlantic currents is being weakened by meltwater flowing from adjacent ice sheets, affecting the global system of ocean currents, and possibly the climates of eastern North America and Western Europe. And the forest systems of the Amazon, of Indonesia and of central Africa are in danger of dieback and of morphing into savannahs, which risks the greenhouse effect intensifying beyond control (Cho 2021). Already extreme weather events, such as floods, droughts, hurricanes, heatwaves and wildfires, are becoming more frequent and more extreme, each of them causing increased fatalities; and climate change is leading to migrations of the vectors of diseases such as malaria and dengue fever to higher altitudes and higher latitudes, and to the migration of environmental refugees who can no longer support themselves in their traditional homes to more benign regions, often by crossing international borders.

These alarming trends present ethical problems in themselves, and also indicate that regional and global systems are under strain, and are some way towards the stage at which tipping points are bound to be reached. Meanwhile, scientists have increasing confidence (and increasing unanimity) in ascribing these trends and tendencies to the global heating that results from human activity of the period since 1850, or, in other words, to anthropogenic climate change.

The reality of climate change needs first to be squarely presented in greater detail, through making clear the problems of global heating and of climate change and related problems. In chapter 2, the findings of climate scientists (including John Houghton) will be presented, which disclose both the comprehensiveness and the urgency of the climate crisis. Two related crises will be presented in chapter 3, those of air pollution and of biodiversity loss. These are problems which share some of the same causes and underline the far-reaching extent of the crisis and at the same time its current impact, particularly on the inhabitants of Earth's towns and cities. These chapters will also serve to show that these problems are in large part anthropogenic, problems (that is) caused by humanity.

The following two chapters concern the ethics of our treatment of contemporary human societies, of future generations, and of fellow species. There is a long tradition that studies ethical reasoning, and the resources of this tradition are brought to bear on the issues (and the options) disclosed by scientific findings, as was attempted ten years ago by John Broome (Broome 2012). Relevant issues concern the nature of environmental justice, what difference we can and should make to generations not yet born or conceived, and whether we have duties in the matter of preserving non-human species and their habitats for their future generations and for our own.

As has been mentioned, 'we', the holders of these duties, include both individuals and governments (among others). So these duties have political implications (presented in chapter 6), involving international collaboration and intergovernmental assistance, as well as implications for domestic policies. Climate engineering is another field where political considerations rest ultimately on ethical ones. Ethics will be found to relate to such household matters as installation of solar panels and frequency of flying, and such intergovernmental ones as support for infrastructure adaptation and concerted plans for greenhouse gas mitigation. Nothing less is involved when climate ethics is investigated.

The severity of the problems does not authorize people to despair in face of apparent collapse, as is shown in chapter 7.

The crises should generate neither apathy nor terror, but proportionate concern, combined with hope that, through concerted action, they can be addressed and ameliorated. The well-being of our contemporaries, of our successors, and of the other species with which we share planet Earth can still be retrieved and secured, if this generation heeds its responsibilities and participates in ethical solutions.

Recommended reading

Broome, John. 2012. *Climate Matters: Ethics in a Warming World*. New York and London: W. W. Norton & Company.

Caldecott, Julian. 2022. 'Implications of Earth System Tipping Pathways for Climate Change Mitigation Investment' (working paper, June). Bristol: Schumacher Institute for Sustainable Systems.

Cho, Renée. 2021. 'How Close Are We to Climate Tipping Points?' *State of the Planet*, 11 November. Columbia Climate School. https://news.climate.columbia.edu/2021/11/11/how -close-are-we-to-climate-tipping-points

Houghton, John. 2015. *Global Warming: The Complete Briefing*, 5th edn. Cambridge: Cambridge University Press.

Singer, Peter. 2002. *One World: The Ethics of Globalization*. New Haven, CT and London: Yale University Press.

Further reading

BBC. 2021. 'COP26: What Was Agreed at the Glasgow Climate Conference?' BBC News: Science & Environment, 15 November. https://www.bbc.co.uk/news/science-environment -56901261

Intergovernmental Panel on Climate Change (IPCC). 2013. 'Summary for Policymakers', in T. F. Stocker et al. (eds), *Climate Change 2013: The Physical Science Basis: Contribution of Working Group I to the Fifth Assessment Report of the Intergovernmental Panel on Climate Change*. Cambridge: Cambridge University Press, 3–29. https://www .ipcc.ch/site/assets/uploads/2018/02/WG1AR5_all_final.pdf

Singer, Peter. 2023. *Ethics in the Real World: Essays on Things That Matter*. Princeton, NJ and Oxford: Princeton University Press.

2
The Science of Climate Change

The greenhouse effect

The Earth is warmed by radiation from the sun. Some of this radiation is absorbed, but much is radiated back. The atmosphere prevents some of this radiation from escaping (infrared radiation included), acting like a blanket. If there were no atmosphere, the Earth would be much colder. This entrapment of radiation is known as 'the greenhouse effect'.

Within a greenhouse, some of the radiation from the plants and their surroundings is trapped and reflected back by the glass. This makes the greenhouse warmer than the ambient environment. The Earth's atmosphere acts in a similar way, trapping heat and making our planet warm enough to be hospitable to life.

But the greenhouse effect is getting stronger, and the atmosphere is becoming warmer in consequence. This change is associated with an increase of various 'greenhouse gases', of which carbon dioxide is the foremost by volume. For the period 1000–1750 CE (and for much of the time since the last ice age), the concentration of carbon dioxide remained at around 280 parts per million (ppm). But by 2000 CE, it had reached 368 ppm, and it currently stands at well over 400 ppm (Houghton 2015: 70; according to Caldecott and his distinguished sources, at 421 ppm in 2022 (Caldecott 2022a: 3: UCSD and SIO

2022). Across the same period (since the year 1750), average temperatures have risen by over 1°C (Caldecott 2022a: 72, Fig. 4.4), most of this increase being attributable to human activity. That may not sound much, but superficially small increases can have huge impacts, as will shortly be seen. (A shorter version of Caldecott's essay was published later in 2022: see Caldecott 2022b.)

Nor is carbon dioxide the only greenhouse gas. Methane (a gas generated by the decay of organic matter) is another, and one that is much more potent, volume for volume, at that. The average concentration of methane from 1000 to 1750 CE was 700 parts per billion (ppb), but this had increased to 1,750 ppb in the year 2000, and to 1,809 ppb in 2012 (Houghton 2015: 70). According to Caldecott and Tollefson, this average level reached over 1,900 ppb in 2022 (Caldecott 2022a: 3; Tollefson 2022), after a record increase during 2020–2021; this total amounts to 262 per cent of the pre-industrial level (Horton 2022). One human contribution to methane levels is made through the manufacture of plastics, for microplastics (the tiny fragments into which plastic bottles and wrappings are shredded by the action of seas and oceans) emit methane as they break down (Winters 2022: 11). Surprisingly, reservoirs contributed 5.2 per cent of global anthropogenic methane emissions in 2020 (Soued et al. 2022). Methane is particularly significant (among other reasons) because 'vast amounts of methane . . . are trapped in permafrost on land and under the sea around the Arctic' (Broome 2012: 75). If the warming of the atmosphere causes much of this methane to escape, then the greenhouse effect will be hugely accelerated.

A third significant greenhouse gas is nitrous oxide, generated through the use of nitrogen fertilizers and from the use of diesel in the engines of cars and lorries. From 1000 to 1750, its concentration in the atmosphere was 270 ppb; by 2012, this had risen to 325 ppb (Houghton 2015: 70), and, by 2022, to 335 ppb (Carrington 2022: 4). As we shall see in the next chapter, as well as being a potent greenhouse gas, nitrous oxide is also a source of air pollution and a widespread direct threat to human health.

Water vapour is also a greenhouse gas, but no one suggests that its presence is substantially due to human activity or significantly subject to human control. However, there are other greenhouse gases, such as CFCs (chloro-fluorocarbons) and HCFCs (hydrochlorofluorocarbons), used till recently as refrigerants, which are greenhouse gases and were found to be destroying stratospheric ozone. That layer of ozone is vital for preserving most life on Earth from ultraviolet radiation. So the world's nations agreed to ban the sale and purchase of CFCs and HCFCs through the Montreal Protocol of 1987 and subsequent agreements (Broome 2012). Since then, the ozone layer has been recovering, although recent wildfires (of the 2020s) seem to be casting this recovery into question.

However, HFCs (hydrofluorocarbons), which were introduced as a substitute for these banned substances, were found to be another greenhouse gas and equally harmful to the ozone layer as well. So they too were banned in an agreement made at Kigali in 2016; if this is implemented, then HFCs will cease to be emitted, and, come to that, to be generated. The Montreal and Kigali agreements supply grounds for hope that international collaboration over environmental threats has proved achievable, and could prove so again, serving to rebut the view that planetary collapse is inevitable.

Ozone at low altitudes (largely a result of human-generated pollution) is yet another greenhouse gas itself, and has almost doubled since pre-industrial levels (Broome 2012: 24). It was estimated in 2012 that the combined impact of all the greenhouse gases, other than carbon dioxide (and water vapour), was equivalent to 11.5 per cent of the (then) level of carbon dioxide. That would bring the total of carbon-equivalent greenhouse gases now (2023) to around 450 ppm.

All the various greenhouse gases absorb warming infrared radiation, and this well explains the phenomenon of global warming, which is itself readily observable from rising average air temperatures and has been found to be affecting the oceans to a depth of at least 3 km (Broome 2012: 27). This warming also explains most of the melting of ice caps and glaciers. This in turn explains the observed rises in sea levels, which amounted to 17 cm in the twentieth century. Sea levels continue to rise at an average of 3 cm per decade (ibid.).

There is plenty of empirical evidence for all this. For example, the increases of greenhouse gases in the atmosphere are evidenced by the proportions present in air bubbles trapped in ancient ice cores from Antarctica, and comparisons of these with average proportions in the current atmosphere. Nor is there significant room for doubt that these increases are due to human activity in the period from the Industrial Revolution onwards. The various greenhouse gases have resulted from emissions from domestic and industrial fires, and from the exhaust pipes of vehicles, ships and aircraft. As Broome adds, atmospheric carbon consists of a mixture of different carbon isotopes, and study of these isotopes shows that a sizeable proportion of this carbon comes from human sources (or is anthropogenic) (Broome 2012: 26).

Other theories have been advanced to explain global warming (as it used to be called), or global heating (an improved description). For example, sunspot activity has been presented as possibly explaining variations in average planetary temperatures across recent years. Perhaps this solar activity really makes a (very) small contribution. But scientific studies show that solar output since at least 1978 has been virtually constant, with fluctuations of only 0.1 per cent between its maximum and minimum extent (Houghton 2015: 158–9). The anthropogenic component turns out to be enormously larger, with sunspot activity playing a relatively tiny supporting role (see Houghton's diagrams, ibid.: 72). The alternating southern Pacific weather systems of Il Niño and La Niña complicate matters, and have the effect that some years are not as hot as their predecessor; but this does not detract from the anthropogenic greenhouse gas theory being substantially vindicated. Indeed, no theory explains the data as well as the anthropogenic one (Broome 2012: 28).

Scientific knowledge of the climate is summarized every five or six years in the reports of the Intergovernmental Panel on Climate Change (IPCC), a body which consults virtually the whole community of climate scientists. These reports give access to the latest scientific findings and conclusions to non-scientists such as Broome and myself. (Broome's conclusion, expressed at the end of the last paragraph, was closely based on IPCC reports.) A reference is given in the full list of references for the section of the IPCC's Sixth Assessment Report that relates to Europe and its climate problems (Bednar-Friedl, Biesbroek and Schmidt 2022), and another to the IPCC report presented in March 2023, urging rapid action to counter increases in greenhouse gas emissions (IPCC 2023). Non-scientists are also fortunate to have previous IPCC findings explained and summarized in John Houghton's *Global Warming: The Complete Briefing* (2015). Houghton is the retired chair of one of the IPCC panels, and his book supplies a comprehensive and up-to-date overview of this whole field of science. While there remains a theoretical possibility that the IPCC stance could be undermined (on climate scepticism, see Lejano and Nero 2020), the evidence that it adduces for global warming, its causes and its threats is close to conclusive.

It should be added that in 2009 a group of climate scientists published in *Nature* a paper showing that, for a 50 per cent chance of avoiding a 2°C rise in temperatures above pre-industrial levels, humanity is limited to emitting (from 1750, the dawn of the Industrial Revolution) just one trillion tonnes of carbon. But, by 2009, more than 55 per cent of this total had been emitted already, and at the rates of emission then current, this 'carbon budget' stood to be used up completely by February 2044 (Meinshausen et al. 2009). Yet if the goal is to avoid a rise of 1.5°C, as is argued below to be necessary, then this carbon budget is considerably reduced. How to share out the remaining carbon budget is both an ethical and a political issue, to which I shall be returning in later chapters.

Extreme weather and related impacts of climate change

While the impacts of global heating are likely to last for hundreds of thousands of years, many are being experienced already. There are widespread impacts on ecosystems, endangered species and their habitats (see chapter 3). But impacts of a more immediate kind are being felt in the form of far more frequent and far more intense storms and hurricanes, floods, droughts, wildfires and heatwaves.

In addition to these impacts, sea levels are continually rising. Across the twentieth century, they rose by around 17 cm, and they continue to rise (yet faster) by over 3 mm per year (Houghton 2015: 76: see also Bednar-Friedl, Biesbroek and Schmidt 2022). These rises put at risk all coastal cities and settlements, and endanger the very existence of many small island states such as Tuvalu, Kiribati and Mauritius, and many of the islands of states like Fiji. Unless global heating is limited to 1.5°C, ice caps and glaciers will continue to melt, and the rise of ocean levels will mean that these states will lose all or most of their territories.

The increasing temperatures and carbon-based acidity of oceanic waters are also causing coral reefs to bleach, with the loss of myriads of creatures that depend on them. We return to this topic in chapter 3 in connection with biodiversity loss.

Yet it is not only the Pacific and Indian Oceans that are affected by climate change. The higher temperatures of the waters of the Caribbean Sea and the Atlantic Ocean generate hurricanes of increased ferocity and frequency, wreaking destruction on Caribbean islands and the southern and eastern parts of the United States. New Orleans has already been put at risk, protected though it is by vulnerable levees to hold back flood water, while many other cities stand to be affected by flooding and storm damage.

Unprecedented floods have also affected Bangladesh (where the rivers Ganges and Brahmaputra converge and meet the sea). Quite apart from the resulting fatalities, recent flooding has increased the salinity of 53 per cent of farmland (McAllister 2022). Floods have also affected parts of Europe, resulting in over 200 deaths in July 2021 (Belcher 2022). In the United Kingdom, February 2020 was the wettest February on record, and storm Ciara brought a month's worth of rainfall across parts of West Yorkshire in just 18 hours, leading to widespread flooding (Henderson 2022). Flooding along the middle reaches of the River Severn has been widely reported in the press in several recent years. But far worse has been the situation in Pakistan, where, according to BBC reports, one-third of the country was covered in flood waters in late August 2022.

Meanwhile, the spread of wildfires is threatening the lives as well as the properties of many, as was widely reported in 2021 from areas of western Canada and the United States. When hurricanes, wildfires and floods are taken into account, polling conducted by Data for Progress discloses that as many as 47 per cent of likely American voters are either 'somewhat concerned' or 'very concerned' about being displaced from their homes through an extreme weather event (Data for Progress 2022).

Other continents have also been affected by wildfires, with serious summer fires in late 2021 and early 2022 in Western Australia, South Australia and Victoria, and dozens killed in the larger fires of 2019–2020. These fires triggered algal blooms larger than Australia itself in the distant Southern Ocean, and sent smoke visible from space hundreds of miles out eastwards over the Pacific (Gramling 2021).

In May 2022, heatwaves struck India and Pakistan, with temperatures exceeding 50°C, and causing 90 deaths; these events are calculated to have become a hundred times more likely due to climate change. The same expert reports that in 2021, during another heatwave, a temperature of 49.6°C was recorded at Lytton in Canada (Belcher 2022), while 47°C was recorded in July 2022 in Portugal (Yaron 2022). Even in the normally temperate United Kingdom, the summer heatwave of 2020 was the most significant heatwave in the previous sixty years, and led to over 2,500 excess deaths across the country (Henderson 2022).

These lines were originally drafted during the heatwave of mid-July 2022, with a red alert (betokening danger of death) in force across large areas of England, and an amber warning (indicating danger to health) across the whole of England and Wales and large areas of southern and central Scotland. In parts of England, temperatures in excess of 40°C were realized as forecasted, together with associated wildfires (BBC 2022a). But, as Kenny Stancil writes, 'the United States, China and parts of Africa and the Middle East are also suffering from heatwaves and wildfires, which climate scientists have long warned will increase in frequency and severity as a result of unmitigated greenhouse gas pollution' (Stancil 2022: 2).

Meanwhile, UNICEF reported that another kind of severe weather event, drought, was in 2022 being caused by heatwaves and lack of rainfall in parts of Somalia, Ethiopia and Kenya, resulting in severe malnutrition there, particularly among children (UNICEF 2022). Drought was also widespread in the southern and western United States, as was attested by the National Drought Mitigation Center at the University of Nebraska, with eleven states suffering from historic levels of extreme drought (National Drought Mitigation Center 2022). News reports also convey that the heatwave of July 2022 generated a drought in Spain, while drought in northern Italy was hampering the production there of olive oil and pasta. The French government described the drought of the summer of 2022 as the worst France had ever had (BBC 2022b). Even in Britain, BBC news reports conveyed that drought conditions in summer 2022 led to a ban on fishing in many Welsh rivers and to unusually low water levels in several English reservoirs. But the impacts of droughts in Britain are trivial when compared with those in Africa.

Some of the severe weather events described above called for immediate attention by way of action to rescue and cure victims of malnutrition, to fight fires, and to rescue and re-house those desperately clinging to life surrounded by flood water. For the longer term, adaptation is needed to equip vulnerable areas with infrastructure such as flood defences and early warning systems to prevent human and animal suffering from extreme climate events becoming recurrent. As António Guterres, the UN Secretary-General, recently said, 'we must treat adaptation with the urgency it needs. One in three people lack early warning systems. People in Africa, South Asia and South America are fifteen times more likely to die from extreme weather events. This great injustice cannot persist' (Stancil 2022). (We shall come to issues of justice in chapters 4 and 5. Early warning systems will be returned to later in the present chapter.)

A problem associated with extreme weather events is that of climate refugees. 'Climate refugees' are defined as people who have been forced to leave their traditional habitat, temporarily or permanently, because of marked environmental disruption. According to the United Nations High Commissioner for Refugees (UNHCR), since 2008 an annual average of 21.5 million people have been forcibly displaced from their homelands by weather-related events, such as floods, storms, wildfires and extreme temperatures. These numbers are expected to increase in coming decades; one forecast (from the Sydney-based think tank, the Institute for Economics and Peace or IEP) predicts that 1.2 billion people could be displaced by 2050 by climate change and natural disasters. This figure includes people at high risk from rising sea levels, whose number has increased over the past thirty years from 160 million to 260 million. There is a strong correlation between the countries most vulnerable to climate change and those experiencing conflict or violence. A leading example is Syria, where around 6.6 million people have been forced to leave their country (McAllister 2022). Yet climate refugees lack international recognition, unless they are also direct victims of armed conflict. This is not the place to discuss responsibilities and remedies; this issue will be discussed in chapter 5.

Another impact of climate change is migration of another kind, that of the vectors of diseases such as malaria and dengue fever (including mosquitoes). These vectors thrive better in a warmer world and, as temperatures rise, extend their range both uphill (to higher altitudes) and polewards (to higher latitudes). These changes put at risk human populations that were previously relatively immune from these diseases. Besides tackling the roots of climate change, it has been argued that adaptation should include the intensification of efforts at vector-borne prevention and control (Rocklöv and Dubrow 2020). I shall return to issues such as the nature of adaptation in later chapters.

But there is a pressing need to attend here to the causes of these extreme weather events and their associated spin-offs, causes which almost certainly include anthropogenic climate change. Yet before we can consider what should be done, and who should do it, we need to reflect further on the other impacts of climate change on humanity, and (first) on the systems that climate change is disrupting, partly to understand them better, and partly to discover the timescale within which corrective action is possible.

Tipping points

The issues of natural systems coming under stress and of the tipping points to which they are liable have already been introduced in chapter 1. The stress to which these systems are vulnerable is centrally due to global heating, as emerges when we remember that one of its main manifestations is the melting of ice caps and glaciers. This at once brings in three of the systems, the Arctic ice cap, the Greenland ice cap, and that of the Western Antarctic; it is just as unsurprising as it is regrettable that these systems are endangered.

Scientists who study systems and their tipping points also include further systems which are at risk. For example, the biogeophysicist Timothy Lenton (of the universities of Exeter and East Anglia) cites the Amazon rainforest (widely recognized as endangered by clearances, and at risk of becoming a savannah), boreal forests (dieback of which could cause additional wildfires), the West African monsoon (failure of which is liable to increase droughts in the Sahel) and the Indian summer monsoon (at risk if weather patterns change sufficiently). He further mentions the Atlantic thermohaline circulation (or the system of currents centring on the Atlantic Ocean, of which the Gulf Stream is the best known), and the El Niño/Southern Oscillation (ENSO, the system that affects the currents and airflows of the Pacific and Southern oceans, and is probably vulnerable as global heating affects these oceans and their environments). Lenton goes on to claim that 'Even with the most conservative assumptions, the results [of a recent study] suggest that it is more likely than not that at least one of five tipping points considered [the five most likely] will be passed in a >4°C warmer world' (Lenton 2011: 202). The study referred to here is entitled 'Imprecise Probability Assessment of Tipping Points in the Climate System' (Kriegler et al. 2009).

It is global heating that has put these systems under stress; and if their tipping points are reached, the effects will include (directly or indirectly) significant further increases in global heating, among other impacts. The central aim of Lenton's review is to assess the prospects for early warnings to be given of tipping points being passed, but the issues are too complex to be considered here. Certainly, early warnings could play an important role, particularly if given early enough before the system is irrevocably committed to reaching a tipping point, allowing targeted remedial action (Lenton 2011: 208).

Before we move on, one of Lenton's examples is worth highlighting. Boreal forests include those of West Canada, and their remaining largely intact is important if atmospheric carbon dioxide is to be absorbed by trees, and if their ecosystem is to remain viable for the sake of its other participant species. However, Lenton discloses that the boreal forests of West Canada are currently suffering 'mountain pine beetle infestation', which is destroying many trees. In this context, Lenton proposes, by way of a deliberate effort to counter the reaching of a tipping point, a project of reforestation, implying that such a project might possibly forestall the onset of a tipping point within that ecosystem (Lenton 2011: 208). Here, it is worth observing that where targeted reforestation involves the possibility of forestalling a tipping point that would lead to both greater global heating and considerable ecological damage, a much greater difference to future generations would be made through committing resources to such a project than by tree planting in regions where tipping points are not in prospect. (This observation will be returned to in chapter 4, one of the chapters on ethics.)

Renée Cho, whose definition of 'tipping points' was cited in the previous chapter, has recently attempted to answer the key question 'How close are we to climate tipping points?' (Cho 2021). Regarding the Greenland ice sheet, she reports that scientists have 'speculated that the critical temperature range at which [it] would go into irreversible disintegration is between 0.8°C and 3.2°C of warming above pre-industrial levels'. This alarming but rather imprecise estimate is accompanied by one that is more precise, and no less disconcerting, for the West Atlantic Ice Sheet (WAIS): a recent study has found, as Cho reports, that current world policies, which are heading for almost 3°C of warming, would (if continued) 'result in an abrupt hastening of Antarctic ice loss after 2060'; yet more disconcertingly, she goes on to relate that 'other research suggests that the tipping point for the WAIS lies between 1.5°C and 2.0°C of warming'. As for what is at stake, she adds that 'another new study found that if the WAIS melted, it could raise sea levels three feet more than previous projections of 10.5 ft' (ibid.), implicitly inundating all coastlines globally.

Cho next turns to the Atlantic thermohaline circulation, which, she relates, has slowed by 15 per cent since the 1950s, adding that 'the latest models project' that global heating could weaken it 'by 34 to 45 percent by 2100'. That could, in turn, have a cooling effect on the east coast of the United States and the west coasts of Western Europe, change rainfall patterns, and reduce agriculture in the United Kingdom. She also relates that some studies suggest that the tipping point of this system 'could be reached [with] between 3°C and 5.5°C of warming' (ibid.: 4–5). All this gives strong grounds for avoiding the kind of temperature rise that the world is heading for at present.

Next, Cho turns to the Amazon rainforest, where 'The policies of Brazil's pro-development president, Jair Bolsonaro, have led to widespread clear-cutting and the rate of deforestation in Brazil is the highest since 2008' (ibid.: 5). At this point, she reports that, according to one study, 'If 20–25 percent of the Amazon were deforested, its tipping point could be crossed, and, with reduced rainfall, this vast forest could start to die back, and transition into a savannah, releasing 90 gigatons of CO_2.' One study, she reports, found that dieback would occur if we reach 3°C of warming (ibid.: 5–6). As we shall see in the next chapter, such a transition would also mean a massive loss of biodiversity. Altogether, the state of the Amazon rainforest supplies even stronger grounds for avoiding a 3°C temperature rise.

Cho further alludes to the El Niño/Southern Oscillation. The warming of the oceans could push ENSO past its tipping point, making El Niño-generated events more severe and frequent, and possibly increasing drought in the Amazon (ibid.: 7). Lenton and several distinguished colleagues agree that passing such tipping points is seriously likely and 'too risky to bet against' (Lenton et al. 2019).

One further system is mentioned by Cho, which was not mentioned in Lenton's 2011 article. This relates to the permafrost found in northern hemisphere lands without glaciers, including 'parts of Siberia, Alaska, northern Canada and Tibet', and also in southern hemisphere regions including Patagonia, Antarctica and the Southern Alps of New Zealand (ibid.: 6). A total of 'Fourteen hundred billion tons of carbon are estimated to be frozen in the Arctic permafrost' alone, 'twice as much carbon as is currently in the atmosphere. The melting of the permafrost would release not only vast quantities of CO_2, but also similar amounts of the even more dangerous greenhouse gas, methane' (ibid.: 6). Methane can escape into the sea as hydrates, which can then be thawed by warming seawater. This was already recognized as a significant threat in 2011 (Gardiner 2011: 169–70); but now the threat is beginning to materialize: 'Scientists recently discovered methane leaking from a giant ancient reservoir of methane below the permafrost of the Laptev Sea in the East Siberian Arctic Ocean'. The amount of carbon likely to be released is uncertain, but '2°C of warming could mean the loss of 40 percent of the world's permafrost' (Cho 2021), with huge attendant releases of methane. Fortunately, a scheme was agreed by a hundred countries at COP26 to cut methane emissions by 30 per cent by 2030 (BBC 2021). Nevertheless, the risks from permafrost melting could be among the most threatening of all.

A biosphere commitment point?

But could any one system reaching its tipping point cause other systems to transition or collapse? Or could interactions between the various systems lead any of them to reach tipping points earlier than expected? Cho goes on to answer these questions affirmatively. A recent study, she relates, found that the Arctic systems, ocean systems and the Amazon rainforest could interact, and could do so before average temperature rise reaches 2°C above pre-industrial levels. This could begin with the melting of ice sheets, as 'their critical thresholds are lower' (Cho 2021: 7). The Atlantic system of currents could slow, with more warmer water appearing in the Southern Ocean, leading to more drought in parts of the Amazon; alternatively, changes in Atlantic currents could 'trigger changes in ENSO, leading to a more permanent El Niño state, whose impacts could lower the critical threshold for Amazon dieback' (ibid.: 8). Long timescales could be involved, but the risk remains a significant one.

Julian Caldecott takes up the theme of interacting systems, and also introduces the concept of a system's commitment point, a point or stage at which a system has so long been under stress (anthropogenic stress included) that reaching its tipping point has become inevitable, whatever human interventions may be attempted (Caldecott 2022a: 7). This concept allows him to introduce the notion of a biospheric commitment point (or BCP), a 'moment when enough Earth systems become committed to transformation that the whole biosphere can no longer resist profound system change, regardless of human efforts thereafter' (ibid.: 9; see also Caldecott 2022b: 6). By this stage, remedial measures such as carbon mitigation could no longer prevent a sequence of tipping points being reached, even if they were not actually reached for several more years.

The same author adds that 'On present evidence, this [the BCP] might tentatively be dated to the year 2050 (plus or minus 10 years)' (ibid.). This is not an IPCC estimate, and the word 'might' should be taken seriously. It may be that even systems well on a pathway to a tipping point (in Caldecott's terms, a 'tipping pathway') could be pulled back by concerted and targeted human action. Yet the possibility of a biospheric commitment point cannot be ruled out; nor can its arrival within the current century.

Caldecott's aim is to postpone the BCP, and specifically to advise the government of Denmark on the policies best suited to doing so. Action is required before a BCP is reached, which makes such action more urgent than may have been apparent. In this connection, Caldecott advocates a 'precautionary' approach, an approach, that is, that employs the Precautionary Principle. This ethical principle is introduced and discussed in chapter 4; but it can here be explained that it is concerned with action by state or regional authorities to prevent significant irreversible harms through timely interventions. And, as Caldecott reminds us, the current context is one in which about '35 billion tonnes . . . of CO_2 were emitted globally in 2020', but where, when the other greenhouse gas emissions are added, total emissions were equivalent to 50 billion tonnes (or gigatonnes) of CO_2 (Caldecott 2022a: 10), if not more.

It is next assumed that the mitigation value of climate investments diminishes exponentially up to the point at which a BCP is reached, which seems a reasonable assumption. This granted, then, the value of earlier investments is vastly greater than that of investments made shortly before the BCP (ibid.: 10): an important conclusion when policies and their timing are being considered.

Besides, as Caldecott further remarks, postponing the BCP would be desirable for reasons including gaining time to achieve a 'just transition' to a sustainable world order, as commended in the preamble to the Paris Climate Agreement of 2015 (Caldecott 2022a: 13; Paris Agreement 2015), and to build the foundations of the 'peace with nature' for which the UN Secretary-General has called (Caldecott 2022a: 13; Guterres 2021). As Caldecott concludes, such measures could possibly postpone the BCP indefinitely. 'Some could be delivered through public mobilization, some by private institutions and some by governments' (Caldecott 2022a: 14). These are measures of carbon (and carbon-equivalent) mitigation, and thus central for present purposes. Some examples are now considered.

In this connection, Caldecott explores the relative values of projects for avoiding deforestation (in Indonesian forests), of renewable energy substitution for fossil fuels (in Ethiopia) and of capacity building for renewable energy generation (in South Africa). While tree planting is approximately as cost-effective as wind farming in securing net emission reductions, avoidance of deforestation is found to be much more cost-effective than either. The avoidance of deforestation is more expensive than capacity building in the energy sector, but the latter is held to be less certain in attaining desired outcomes, which are largely beyond the donors' control (Caldecott 2022a: 13). It can be commented that outcomes beyond emission reductions (such as social ones) should also be taken into account and, further, that these comparisons almost certainly fail to take into account the value of the non-human species affected. Yet the value of avoiding deforestation in terms of preserving non-human species would be likely to add further to the overall value of projects of this kind.

These findings take us beyond what IPCC discloses, but are still helpful in showing how important scientific modelling and systems theory can be, whether we grant assumptions such as the prospect of an early BCP or plan to postpone BCP for as long as possible. However, it is time to return to the more widely publicized findings of climate science.

Further impacts of climate change on humanity

Before we return to the various impacts of climate change on humanity, it is worth quoting the verdict of the IPCC report of 2013 on the human contribution to climate change:

Human influence has been detected in warming of the atmosphere and the ocean, in changes in the global water cycle, in reductions of snow and ice, in global mean sea level rise, and in changes in some climate extremes. This evidence for human influence has grown since AR4 [the IPCC report in 2007]. It is *extremely likely* that human influence has been the dominant cause of the observed warming since the mid-twentieth century. (IPCC 2013: 17; see also Houghton 2015: 87)

The closing words concern the dominant causes of the warming observed since 1950, without casting doubt on human activity being the dominant cause of global warming since at least the Industrial Revolution. Greenhouse gases have increased since that revolution, and, as Broome remarks, that is itself a clue about the cause (Broome 2012: 26).

We should also note that this human activity is not the activity of the whole of humanity, but of those human beings and human groups who have emitted more greenhouse gases than our ancient and medieval predecessors. Many people have in fact contributed little or nothing to climate change. But almost everyone has been and is being affected by it, some only trivially, some seriously, and some fatally. (These contrasting facts raise issues of justice: see chapters 4 and 5.)

Impacts include the deaths and injuries that are due to extreme weather events. Not all are caused by climate change (tsunamis for example are not), but the increased frequency and severity of episodes of extreme weather, as predicted by climate scientists, mean that their casualties are mostly victims of climate change and little else.

Damage to ecosystems often has impacts on human beings, such as those reliant on fish for their diet. There are other causes, such as overfishing, but the damage to coral reefs (by temperature rise and ocean acidification) and to mangroves (by storms and rising sea levels) contributes to the falling catches of many fishermen. Damage of this kind is discussed further in chapter 3.

Climate change makes some places (for example, parts of northern Europe) more fertile, with increases in crop yields. But tropical farming is being damaged as temperatures rise, as is farming everywhere by temperature rises of more than a degree or two. Besides, dry areas are becoming drier and wet areas wetter, with increasing droughts and increasing floods that are almost all harmful to farming (Broome 2012: 32). Climate refugees have been mentioned already; many flee because their traditional homelands have become either too drought-ridden or too flooded (in some cases because of rising seawater).

Broome points out that the people of some small islands will have to be evacuated, and that populous river deltas are particularly vulnerable. One study, he reports, estimates that, by 2500, 3.43 million of the people of the combined delta of the Ganges and Brahmaputra rivers will be displaced by rising seas, and a further 4.7 million will become victims of flooding during storm surges, many thousands of them drowning. Globally, nine million people will be displaced from their homes in deltas by 2050 (Broome 2012: 32; Ericson et al. 2006: 63–82).

When people die during famines, malnutrition is widely the underlying cause (Crocker 1996), and malnutrition is often caused by droughts resulting from climate change. Heatwaves, floods and storms, many due to climate change, will kill others. As Broome adds, poor air quality, often associated with climate change, will kill yet others through respiratory diseases: for more discussion of this problem, see chapter 3. The increased range of malaria and dengue fever has already been mentioned; these and other tropical diseases will cause deaths in areas that were immune prior to climate change (Broome 2012: 32).

The World Health Organization estimates that 141,000 deaths in 2004 were attributable to climate change, even though average temperature rise was at that time no more than three-quarters of a degree (Broome 2012: 33; World Health Organization 2009). One report predicts a million deaths per year from 2030 onwards caused by climate change (Broome 2012: 33; Climate Variability Forum 2010).

Broome adds that the *Stern Review* estimates that climate change, if not controlled, will diminish the world's economic production by at least 5 per cent, and perhaps as much as 20 per cent, into the envisageable future (Broome 2012: 33; Stern et al. 2007). This report has received strong criticism from other economists, but this estimate well presents the order of the problem, if not the precise losses to the global economy.

The ongoing and increasing harms due to climate change already make concerted action by individuals, families, corporations and governments indispensable. The findings of IPCC and other scientists about tipping points, and the prospect of a domino-like cascade of tipping points across climate systems, supply a strong case for the urgency of such action. Yet the climate crisis is related to two other crises that need to be taken into account. These are considered in chapter 3.

Recommended reading

Caldecott, Julian, 2022a. 'Implications of Earth System Tipping Pathways for Climate Change Mitigation Investment' (working paper, June). Bristol: Schumacher Institute for Sustainable Systems'.

Caldecott, Julian. 2022b. 'Implications of Earth System Tipping Pathways for Climate Change Mitigation Investment' (working paper). *Discover Sustainability* 3 (8 November): https://link.springer.com/article/10.1007/s43621-022-00105-7

Cho, Renée. 2021. 'How Close Are We to Climate Tipping Points?' *State of the Planet* 11 November. New York: Columbia Climate School. https://news.climate.columbia.edu/2021/11/11/how-close-are-we-to-climate-tipping-points

Crocker, David A. 1996. 'Hunger, Capability and Development', in William Aiken and Hugh LaFollette (eds), *World Hunger and Morality*, 2nd edn. Upper Saddle River, NJ: Prentice-Hall, 211–30.

Houghton, John. 2015. *Global Warming: The Complete Briefing*, 5th edn. Cambridge: Cambridge University Press.

Kriegler, E., Hall, J. W., Held, H., Dawson, R. and Schellnhuber, H. J. 2009. 'Imprecise Probability Assessment of Tipping Points in the Climate System'. *Proceedings of the National Academy of Science USA* 106 (13): 5041–6.

Lenton, Timothy M., Rockström, Johan, Gaffney, Owen, et al. 2019. 'Climate Tipping Points – Too Risky to Bet Against'. *Nature* 575: 592–5. https://doi.org/10.1038/d41586-019-03595-0

McAllister, Sean. 2022. 'There Could Be 1.2 Billion Climate Refugees by 2050'. Zürich Insurance Group. https://www.zurich.com/en/media/magazine/2022/there-could-be-1-2-billion-climate-refugees-by-2050-here-s-what-you-need-to-know

Rocklöv, Joacim and Dubrow, Robert. 2020. 'Climate Change: An Enduring Challenge for Vector-Borne Disease Prevention and Control', *Nature Immunology* 21: 479–83, 20 April, https://doi.org/10.1038/s41590-020-0648-y

Stern, Nicholas et al. 2007. *The Economics of Climate Change: The Stern Review*. Cambridge: Cambridge University Press, ch. 3.

World Health Organization. 2009. *Global Health Risks: Mortality and Burden of Disease Attributable to Selected Major Risks*. https://apps.who.int/iris/handle/10665/44203

Further reading

BBC. 2021. 'COP26: What Was Agreed at the Glasgow Climate Conference?' BBC News: Science & Environment, 15 November. https://www.bbc.co.uk/news/science-environment-56901261

Carrington, Damian. 2022. 'World is Coming Close to Irreversible Change, Say Climate Experts'. *Guardian*, 28 October, 4.

Gardiner, Stephen M. 2011. *A Perfect Moral Storm*. Oxford: Oxford University Press.

Guterres, António. 2021. 'UN Secretary-General's Foreword', in *Making Peace with Nature: A Scientific Blueprint to Tackle the Climate, Biodiversity and Pollution Emergencies*. Nairobi: United Nations Environmental Programme.

Intergovernmental Panel on Climate Change (IPCC). 2023. *Climate Change 2022: Impacts, Adaptation and Vulnerability*. https://www.ipcc.ch/report/sixth-assessment-report-working-group-ii/

Lenton, Timothy M. 2011. 'Early Warning of Climate Tipping Points'. *Nature Climate Change* 1 (July): 201–9.

Meinshausen, Malte et al. 2009. 'Greenhouse Gas Emission Targets for Limiting Global Warming to 2°C'. *Nature* 458 (30 April): 1158–62.

Paris Agreement. 2015. Preamble. https://unfccc.int/files /essential_background/convention/application/pdf/english _paris_agreement.pdf

Tollefson, J. 2022, 'Scientists Raise Alarm over "Dangerously Fast" Growth in Atmospheric Methane'. *Nature*, 8 February. https://doi.org/10.1038/d41586-022-00312-2

UNICEF. 2022. 'At Least 10 Million Children Face Severe Drought in the Horn of Africa', 22 April. https://www.unicef .org.uk

3

Related Crises: Biodiversity Loss and Air Pollution

Tipping points and mass extinctions

Tipping points have at times befallen life on Earth, as well as Earth's major systems. As the scientist Anthony D. Barnosky writes, one happened when 'the last ice age suddenly gave way to warmer times' (and, we could add, to greener times) about twelve thousand years ago. Another occurred 'when what used to be verdant wetlands abruptly turned into the Sahara Desert some four thousand years ago' (Barnosky 2015, updated 2017). Barnosky hopes for a worldwide social tipping point (adjusting the sense of this phrase slightly) to more benign relations between humanity and nature, with greatly reduced greenhouse gas emissions. But for the moment the point is that living systems can undergo tipping points just like systems of ice sheets and of ocean circulation. This was already apparent in chapter 2 with regard to the Amazon rainforest and to the boreal forests.

These remarks of Barnosky continue to comply with the definition of 'tipping point' subsequently supplied by Cho, and cited in chapter 1: 'the point at which small changes become significant enough to cause a larger, more critical change that can be abrupt, irreversible and lead to cascading effects' (Cho 2021). Social tipping points, it should be admitted, might fall short of being irreversible, but steps could be taken, through widespread programmes of environmental education, to make the reversal of a social tipping point unlikely. In any case, Cho's definition only says that tipping points *can* be irreversible, not that they are always so.

But, as Barnosky recognizes (see the title of his 2014 book, *Dodging Extinction: Power, Food, Money and the Future of Life on Earth*), further historical tipping points have taken the form of mass extinctions. Five historical mass extinctions are recognized by science, together with the possibility that a sixth is already under way, as has been argued by Elizabeth Kolbert in *The Sixth Extinction: An Unnatural History* (Kolbert 2014). The five have included the famous K-T event, otherwise known as the Cretaceous-Tertiary Extinction Event of 65.5 million years ago, when three-quarters of all plant and animal species, including the dinosaurs, became extinct. ('K' is the recognized symbol for Cretaceous, to avoid confusion with the Carboniferous Age.) This was not the largest ever historical extinction event, but it was in some ways the most significant because it allowed mammals (and eventually human beings) to become the dominant vertebrates on land (*Encyclopaedia Britannica* 2020). Rupert Read comments that the sixth extinction puts at risk the biosphere (in his language, 'Gaia'), adding that, 'According to the authoritative work done by the Stockholm Resilience Institute, the breaching of planetary boundaries by humanity is even worse vis-à-vis biodiversity/extinction than it is vis-à-vis climate!', citing here the Attenborough and Rockström film, *Breaking Boundaries: The Science of Our Planet* (Read 2022).

According to the World Wide Fund for Nature (WWF), the sixth mass extinction, unlike its predecessors, is caused by human activity, 'primarily . . . the unsustainable use of land, water and energy use, and climate change'. 'Thirty per cent of all land' that used to sustain biodiversity (biological diversity, that is, or diversity of species, sub-species and habitats) 'has been converted for food production. Agriculture is also responsible for 80 per cent of deforestation, and accounts for 70 per cent of the planet's freshwater use, devastating the species that inhabit those places by significantly altering their habitats' (WWF 2022: 2). Further, 'unsustainable food production and consumption are significant contributors to greenhouse gas emissions that are causing atmospheric temperatures to rise' (ibid.: 2), with all the impacts that have been specified in chapter 2. These include severe droughts and intense storms that make the habitats of many species 'inhospitable' (ibid.: 2).

Humanity obviously needs agriculture and food production, but not of unsustainable kinds (or future generations will suffer); and the deforestation of tropical forests is proving disastrous not only for the climate, but also for the species that inhabit them. Some people are untroubled by the loss of species and habitats; reasons for caring about their good will be presented in chapter 5. But, as the WWF remarks, wild species are in any case interconnected, and the decline (or extinction) of one or two can undermine the ecosystem to which they belong, reducing its ability to provide human beings with clean air, clean water and healthy soils (ibid.: 3), even though some species are not as crucial as others. And when local ecosystems break down, larger systems (such as those of the Amazon, Congo or Borneo rainforests) come under threat, which brings an increased risk of a cascade of major systems globally. Accordingly, human well-being is very much at stake.

The World Wide Fund for Nature goes on to relate that the current rate of species extinction is estimated as between 1,000 and 10,000 times higher than natural extinction rates, the rate of species extinctions 'that would occur if we humans were not around' (ibid.: 3). Extinctions are a normal component of the evolutionary process, but current rates of population decline (particularly among birds and insects) and of species extinctions are high enough to threaten those 'ecosystem services' that support life on Earth (including human life), in the forms of a stable climate, predictable rainfall patterns, and 'productive farmland and fisheries' (ibid.: 3: for another view, see Newman, Varner and Linquist 2017). Neither biodiversity loss nor extinctions can be comprehensively conveyed here, but enough statistics and examples can be supplied to disclose the extreme seriousness of this problem.

Biodiversity loss: extent and examples

According to the *Living Planet Report*, the world has seen an average 68 per cent drop in mammal, bird, fish, reptile and amphibian populations since 1970, with population losses of nearly 3 billion birds over the same period (Almond, Grooten and Petersen 2020; see also Mulhern 2020). Biodiversity loss has reached an almost incredible 94 per cent (population-wise rather than species-wise) in Latin America and the Caribbean, and 65 per cent in Africa, again since 1970 (WWF 2022).

Insects, which E. O. Wilson once described as 'the little things that run the world', are severely declining in population numbers on an ongoing basis, most evident in Western Europe and North America, where intensive agriculture was pioneered. If such changes of land use and anthropogenic disturbance spread worldwide, there is reason to expect this decline to become global (Almond, Grooten and Petersen 2020). Insects, it hardly needs to be mentioned, include bees and other pollinators, and that the long-standing human reliance on insect pollination can no longer be depended on (see pp. 33–6). Efforts to restore wildflower meadows could help reverse the decline of insects like wild bees and moths.

The *Living Planet Report* estimates, on the basis of available data, that almost 90 per cent of the world's wetlands have been lost since 1700, and that the 944 monitored species of freshwater species (mammals, birds, amphibians, reptiles and fish) have declined as populations by an average of 84 per cent since 1970, a decline equivalent to 4 per cent per year over that period. The largest declines have affected the larger river species, including megafauna (species that grow to more than 40 kg), such as river dolphins, otters, beavers and hippos (Almond, Grooten and Petersen 2020).

As for fish, the Center for Biological Diversity relates that in the summer of 2021 'scorching temperatures caused mass fish die-offs' and that in the summer of 2022 'wildlife [was] faring no better'. They add that climate change is threatening Mount Rainier white-tailed ptarmigans, salmon, American pikas, freshwater mussels and many other (American) species. Their staff member Noah Greenwald warns that if humanity does not change course, the result could be extinctions 'at a rate our world hasn't seen in at least 66 million years', a period that includes the K-T event (Center for Biological Diversity 2022).

In the same despatch, the Center for Biological Diversity reports that monarch butterflies have been placed on the International Union for Conservation of Nature red list of threatened species. By August 2022, of 142,577 species assessed, 40,084 species were listed as threatened (28 per cent of all assessed species), species ranging from African elephants to dragonflies. This list is a self-proclaimed 'barometer of life', a critical indicator of the world's biodiversity. Some species are 'on the path to recovery', including four commercially fished species of tuna; this is attributed to 'the enforcement of regional fishing quotas over the last decade'. Also, the European bison, Europe's largest land mammal, has moved up from 'Vulnerable' to 'Near Threatened', thanks to continued conservation efforts. But the predominant picture is of increasing numbers of creatures becoming 'critically endangered', such as the dawn jewel damselfly, due to the ongoing destruction of wetlands, and more widely as a result of the droughts and wildfires attendant on climate change (IUCN 2022a, 2022b).

According to *Science Daily*, the latest estimate of the number of (non-bacterial) species on Earth is 8.7 million, with 6.5 million species on land and 2.2 million in the oceans. This figure was announced by the Census of Marine Life on 24 August 2011. The same study adds that 86 per cent of all species on land and 91 per cent of those in the seas have yet to be discovered, described and catalogued, and that 'many species may vanish before we even know of their existence, of their unique niche and function in ecosystems, and of their potential contribution to improved human well-being' (*Science Daily* 2011). (Nor, obviously, can we know of their actual contribution to human well-being while they existed undiscovered.) It follows that fewer than two million species have been identified, and that many of the others are or will be victims of current and future extinctions. By the same token, it also follows that the number of species listed on the International Union for Conservation of Nature (IUCN) red list is far smaller than the full total of threatened species.

Pollinators under threat

This brings us back to the recent placing of monarch butterflies (*danaus plexippus*) on the IUCN red list, for it is far more feasible to discuss species known to be endangered than the many others that are unknown. For one thing, we can trace the impact on monarchs of changes in their habitats; for another, we need to grasp some of the impacts of their decline on other species. These highly attractive butterflies have declined by at least 65 per cent in two decades. They migrate a thousand miles every year, many from the north-west of the United States to California; and on the way they act as the pollinators of various crops. This makes their potential disappearance a serious matter not only for admirers of their beautiful orange and black patterns, but also because of their economic significance. Yet we have apparently now lost

90 per cent of their erstwhile population (Friends of the Earth 2022).

A large part of the problem is the evolved dependence of monarch butterfly caterpillars on a plant called 'milkweed'; yet 850 million milkweed plants have been lost, and have disappeared in particular in several of the states where monarchs used to breed. This massive disappearance of milkweed plants is due in large part to the use of the toxic herbicide glyphosate, which farmers often spread on their fields to eradicate insects that they regard as pests (ibid.), as well as to remove weeds. This is far from the first time that pesticides have been found to be producing unintended adverse effects: remember how Rachel Carson publicized the spread of DDT, used in the northern hemisphere to counter insects such as mosquitoes, to the flesh of the faraway penguins of Antarctica (Carson 2000 [1962]), thus triggering the environmentalist movement. So part of the reason for the alarming decline of monarch butterflies is chemical pollution from the agrochemical industry. Either more selective pesticides should be used instead, or resort should be had to organic pesticides (pyrethrum being one of these).

The decline of monarchs west of the Rockies is estimated by the IUCN at as great as 99 per cent. Significantly, the changes to which this collapse is ascribed include habitat destruction and climate change as well as pesticides. (It should be remembered that some habitat destruction is due to pesticides, and some of the rest to the droughts and desertification that result from climate change.) The IUCN adds that the migrating populations of monarchs are 'less than half the size they need to be to avoid extinction' (Center for Biological Diversity 2022b). Hence their being listed as endangered.

A far more prolific pollinator is the honeybee (*apis mellifera*). The honeybee is one of 25,000 bee species, but, besides making honey, usually in human-made hives, its central modern role is as a pollinator of some third of all our farmed crops, including 'almonds, peaches, soybeans, apples, pears, cherries, a variety of berries, melons, cucumbers, nuts, onions, carrots, broccoli, sunflowers, oranges, avocados, alfafa . . . and cotton' (Mathews 2010: 358). Over a million hives are moved around in lorries so that the bees arrive at the moment the flowering season of almonds (etc.) begins, and are then moved on relentlessly to the next crop, and the next (ibid.: 356).

Most of these crops could, in theory, be pollinated by other species, but 'far less effectively and less readily, especially as many of these insects have already been displaced in the wild by the commercial honeybee and her feral relatives' (Mathews 2010: 358). It appears that 'the solitary blue orchard bee . . . is currently being groomed to join the honeybee as a commercial pollinator', but, because they are 'solitary rather than social, [they] cannot be bred in the numbers needed for commercial pollination on a grand scale' (Mathews 2010: 358; Mims 2009). The other option is to pollinate crops by hand, but such a practice would be too labour-intensive to be economically feasible (Mathews 2010: ibid.).

The search for an alternative arises because hives in many parts of the world are being found empty of workers, with the bees dying from 'colony collapse disorder', first observed in 2007 (Mathews 2010: 357). Modern hives have existed only since the eighteenth century, and have allowed a mutually advantageous relationship between honeybees and humans; but the stresses of commercial pollination practices, involving the breeding of bees in their billions, without regard for apiary life cycles, may play a part in these collapses. Another suggested cause is the varroa destructor mite, a well-known parasite of honeybees; yet there are bee colonies infested with the mite, but not suffering collapse. A further possible cause of this mysterious malady could be the use of neonicotinoids, pesticides known to produce symptoms in other social insects such as termites consistent with those of colony collapse disorder (ibid.). A more recent study confirms that colony collapse disorder spreads infectiously from one hive to its neighbours, and suggests that multiple factors, including habitat loss, pesticides and climate change, could be jointly responsible for this disorder (Zissu 2022).

Mathews suggests that this widespread and ongoing loss of honeybees opens up a gap in the underlying story of the biosphere, the story which makes it possible to find meaning in life; hence the phrase 'planetary collapse disorder' in the title of her article (Mathews 2010). Yet even if we do not go along with this diagnosis of our grief at the loss of honeybees, the disappearance of honeybees represents a significant breakage of the ecological chain that holds the natural world together, granted that honeybees pollinate not only human crops but many of nature's wild flowering plants as well. The loss of this link in nature's chain, and the likelihood that it is caused by some combination of climate change, technology and industrial activity, suggests that other gaps could well open up, putting at risk not only pollination but many other natural processes as well, with growing risks of the unravelling of hitherto reliable natural systems and ecosystems.

Coral reefs

Another lost link in nature's chain could be coral reefs, many of which are bleaching and dying. The corals of these reefs are living creatures and harbour at least 25 per cent of all known marine species. Coral reefs are some of the most biologically diverse ecosystems on the planet; they also benefit millions of people as places of recreation and sources of food, and as buffers of shorelines against damage from storms.

But they are widely under stress from pollution, overfishing, unsustainable coastal development and in particular from climate change and ocean acidification. Warming ocean temperatures, together with increasing concentrations of carbonic acid in the waters of the oceans, are causing corals to bleach and often to die, and these trends are predicted to increase in frequency as well as intensity as temperatures rise further. Corals can recover from bleaching, but only when environmental stress is mitigated. Their loss has impacts on thousands of reef-dependent species. According to the Status of Coral Reefs of the World 2020 Report, 14 per cent of the world's corals were lost between 2009 and 2018. Without drastic action to limit global warming, a further 70–90 per cent of corals could be lost by 2050. Yet these losses are preventable if governments and businesses take responsibility for safeguarding their health. Several of the Sustainable Development Goals of 2015 call for action of this kind, such as Goal 14, which seeks to preserve 'Life Below Water' (UNEP 2021).

An intergovernmental conference on the protection of the marine biological diversity of waters beyond national jurisdictions (often called 'the high seas') took place in New York (August 2022). Its aim was to draft a treaty for the preservation of marine life, which would form part of the (legally binding) United Nations Convention on the Law of the Sea (United Nations 2022). Those talks were inconclusive (Helmore 2022), but greater progress was made at the Kunming–Montreal Biodiversity Conference of December 2022 (see chapter 6). Then, in March 2023, agreement was reached on a High Seas Treaty to preserve the marine life of 30 per cent of the oceans (BBC 2023a). Besides complying with such international measures, local and national authorities will need to take related action to preserve the marine life of areas situated within their territorial waters.

Deforestation

Rainforests used once to cover 15 per cent of the Earth's land area. Yet by 2022 they covered less than 3 per cent. Only 2.4 million sq. miles are left. A large part of this destruction has taken place during the last fifty years, with around 50,000 sq. miles of tropical rainforest being lost each year. And the rate of destruction is currently increasing. Almost 800 sq. km of forest were felled during the first three months of 2020, an increase of 51 per cent over the same period of 2019 (YPTE 2022: 1). However, the rate of destruction was even higher in the 1980s, when the annual rate was 15.4 million hectares (Hammond 1994), compared with 6 million hectares at present (YPTE 2022: 1).

In West Africa, almost 90 per cent of the rainforest has been destroyed, and now logging has spread to the rainforest of Central Africa. It is estimated that between 70 per cent and 80 per cent of the logging taking place in Brazil and Indonesia is illegal. Trees are cut down not only for timber, but also for wood pulp, used to make products such as paper. In Brazil, 80 per cent of deforested land is used for cattle ranching. Such deforestation is the country's largest source of carbon emissions; additionally the cattle produce methane, which is 26 times as effective a greenhouse gas as carbon dioxide (ibid.).

Forests and savannahs in Brazil are also cleared to grow soya, a high-protein foodstuff used to feed cattle, particularly in Europe and China. The land is initially productive, but this lasts only as long as the topsoil remains intact. At that stage, fertilizers and pesticides have to be used if soya production is to continue (ibid.).

In Malaysia and Indonesia (which share the island of Borneo between them), many acres have been cleared for the sake of growing palm oil. Malaysia has laws against felling trees right up to river banks, so that corridors are left for the use and migrations of wildlife, and the environmentalist Stuart Pimm campaigns for such connectivity throughout Asia and South America (Schramm 2020); but laws such as these are widely ignored, thus effectively destroying the habitats of many wild creatures. There is also a widespread practice in Indonesia and Malaysia of burning peatlands, again to grow palm oil. But though peatlands occupy only 0.1 per cent of the planetary land surface, their burning contributes as much as 4 per cent to greenhouse gases (ibid.).

Yet rainforests, some of which have remained intact for the last 65 million years, support the greatest diversity of living organisms on Earth, housing an estimated 50 per cent of all species. Whereas temperate forests are dominated by half a dozen species of trees, which make up 90 per cent of the trees in the forest, a tropical rainforest can support more than 480 tree species in a single hectare; and a single bush in the Amazon may harbour more species of ants than the entire British Isles (Mongabay 2006). Over 3 million species of all kinds live in the Amazon rainforest alone, with 2,500 tree species, one-third of all the tropical trees that there are on Earth. There are also numerous species of mammals, birds, insects, reptiles and amphibians (Thomson 2020).

The felling of forests and the related widespread loss of species raises the risk of the whole system of the Amazon reaching a tipping point, with the hydrological system being disrupted and vast areas of the rainforest being replaced by savannah. This transition would involve vast losses of species and thus of biodiversity (Thomson 2020), a prospect discussed in chapter 2 above. Besides the knock-on impacts that this could have on other systems, it would also mean that none of the future generations of the lost species would come into being, with enormous losses both to those species and to the future human generations that could never encounter them.

But we should not forget the boreal forests, such as those of Canada, Scandinavia and Russia, which compose one-third of Earth's forests, and are, like tropical forests, under threat. Boreal forests store between 30 per cent and 40 per cent of the entire planet's land-based carbon. Yet industrial logging removed 25.4 million acres, an area the size of Kentucky, over the twenty years up to 2017. Like tropical rainforests, boreal forests are the home of many indigenous peoples, who depend on the forests for the preservation of their cultures. But these forests are threatened by a combination of extractive industries and climate change. They are warming twice as fast as other parts of the globe, and are endangered by the increasing incidence of wildfires. Scientists fear that their destruction may be approaching a tipping point and thus a point of no return (Ramirez 2021).

Temperate forests too are endangered. Less than 10 per cent is left of the forests of California, Washington and Oregon, largely through intensive logging, which has also contributed to the loss of many plant species. Less than 3 per cent of the original temperate rainforest remains in Australia, and clearances for agriculture have drastically reduced the forests of Europe. The fragmentation of forests has also fragmented the habitats of many mammals, birds and marsupials (Zinni 2018).

An additional reason for deforestation worldwide is the potential for mineral extraction in forested areas. In Indonesia, people often use mercury in their quest for gold, despite its use being illegal, with the result that many rivers are affected by mercury pollution. Brazil has the world's largest iron ore mine, and has rich reserves of zinc, nickel, tin and aluminium; some forest destruction is due to road building to facilitate the extraction of some of these metals (YPTE 2022).

Nevertheless, the UNFCCC Conference of the Parties in 2007 created the programme 'Reducing Emissions from Deforestation and Forest Degradation' (REDD+) 'to combat climate change by rewarding and incentivizing tropical forested countries for reducing emissions from deforestation and forest degradation' (Brown 2019: 262; Kelbessa forthcoming). Despite some misadventures, this programme has the potential to combat both deforestation and climate injustice.

More recently, the leaders of over a hundred countries (with some 85 per cent of the world's forests in their territories) agreed at the Glasgow COP26 Conference of 2021 to halt deforestation by 2030. One of the reasons for this agreement may well have been a desire to prevent biodiversity loss, alongside concern for indigenous peoples and opportunities for eco-tourism, and possibly a fear of transgressing tipping points. Similar agreements in the past have proved ineffective, but this agreement was associated with better prospects of funding (BBC 2021). The subsequent resistance to an international Loss and Damage Fund may have jeopardized the prospects of this agreement being implemented as widely as had been hoped, but the acceptance of such a Fund at COP27 revived these hopes, and was one of its few successes (Varley 2022). This success was followed in the agreement at the Kunming–Montreal Biodiversity Conference (COP15) to preserve 30 per cent of the planet's land surface and 30 per cent of its oceans for biodiversity (Koop, Lam and Zhijian 2022). (We return to these agreements in chapter 6.) Certainly, the very fact that these agreements were possible shows that biodiversity loss is widely recognized even by governments as a global problem, with alarming consequences for humanity as well as other species already, and worse ones in prospect. Read may imaginably be right that it is an even more serious problem than climate change (Read 2022).

The crisis of air pollution

A third global crisis should now be considered, that of air pollution. The pollution of land and of seas and oceans supplies a whole spectrum of problems, from the toxic dumping of industrial wastes (radioactive wastes included) to the oil slicks and plastic pollution that disfigure all our oceans. Yet the pollution of the air of our cities, towns and highways has become a significant threat to human and animal life globally, and governments and other agencies need to be aware of its nature and extent, at the same time as addressing the related problems of climate change and biodiversity loss.

Even people who deny that climate change is caused by humanity cannot cogently deny that the fumes from road and rail traffic and the particles and gases emitted from domestic and industrial fires and from diesel engines are causing air pollution worldwide. What used to appear to be a set of local problems has turned into a serious global problem for human health. This is not (normally) a problem involving weather systems (although weather conditions in places as different as Los Angeles and Beijing sometimes exacerbate it), but one arising everywhere from the social systems of modern transport, modern industry and modern waste disposal.

Writing in the *Guardian*, Fiona Harvey cites the *State of Global Air* report as estimating that in 2019, 1.7 million deaths were due to exposure to airborne particulate matter in as many as 7,239 cities worldwide, with exposures being highest in South and East Asia, Africa and the Middle East. The report adds that 92 per cent of the world's human population lives in areas that exceed one of the key World Health Organization targets for exposure to particulates (Harvey 2022a; Health Effects Institute 2023).

Harvey also discloses that serious levels of nitrogen dioxide are blighting numerous cities in relatively prosperous countries. The world's worst cities for this kind of pollution are Shanghai, Moscow, Tehran and St Petersburg, with Ashgabat (capital of Turkmenistan) and Minsk (capital of Belarus) not far behind. Other seriously affected cities include Cairo, Istanbul and Ho Chi Minh City. In these cities, traffic pollution, often from older fleets of vehicles, seems to be the main source of the problem (Harvey 2022a). But it is not only megacities that are affected; so are other cities, towns and most arterial roads. The World Health Organization estimates that ambient (outdoor) air pollution in both cities and rural areas caused 4.2 million premature deaths worldwide in 2016, with some 91 per cent of these deaths occurring in low- and middle-income countries. It adds that in addition to outdoor air pollution, indoor smoke is a serious health risk for some 2.4 billion people who cook and heat their homes with biomass, kerosene fuels and coal (World

Health Organization 2021b). Meanwhile, in the United Kingdom, the Royal College of Physicians (RCP) estimated in 2017 that approximately 40,000 premature deaths and over 20,000 hospital admissions could be attributed to air pollution every year.

As for the effects of these pollutants, the British government explains that nitrogen dioxide 'irritates the airways of the lungs, increasing the symptoms of those suffering from lung diseases', while 'fine particles can be carried deep into the lungs, where they can cause inflammation and a worsening of heart and lung disease' (Defra 2022). In case this might possibly seem a tolerable level of harm and disease, it is worth highlighting a study conducted by US researchers in Canada, one of the cleanest, least polluted countries on Earth. Despite the relatively clean air of Canada, this study found that nearly 8,000 Canadians were dying prematurely each year from outdoor air pollution, and that even in the cleanest areas people were experiencing an adverse impact on their health (Fuller 2022).

The Health Effects Institute funded two other reports, one focusing on 60 million people in the United States and the other on 27 million people in Europe. All three reports reached similar conclusions, namely that there is no safe lower limit for air quality. Hence governments should not focus on setting targets, as if those levels of pollution would be acceptable, but should instead focus on continued reductions year by year into the indefinite future (Fuller 2022: 2).

Fuller further discloses that, in the summer of 2022, a UK review warned that air pollution contributes to dementia, and also that a US review highlighted how asthma can start through exposure to air pollution from traffic. He adds that, although the United Kingdom and several other European countries are committed to reducing both average particle pollution and total air pollution, the evidence of all these studies 'underlines the need for action to improve air pollution everywhere, and especially in places where young and vulnerable people are liable to be affected' (Fuller 2022: 2). Subsequently, in early September 2022, BBC news bulletins reported that scientists have discovered the way in which air pollution is capable of contributing to the cells of human bodies becoming cancerous.

Nor is it human beings only that are impacted by air pollution. BBC Radio reported, during the *World at One* programme of 31 January 2023, that the black kites, for which the city of Delhi is well known, often fall out of the sky because of air pollution there. Other urban species are likely to be similarly affected (BBC 2023b). These wider impacts underline that in air pollution we have yet another crisis.

Belated legal recognition of air pollution in Britain

Many people must have died from air pollution in Britain during the twentieth century. I can remember walking to school through the streets of Watford in 1952 through smog (a combination of sulphurous smoke and fog) too thick to allow people to see the way more than ten yards in front of them, which extended throughout the Greater London area (an area of which the town of Watford was at the fringe). There were many casualties at that time, whose deaths will have been ascribed to conditions like asthma or pneumonia. Fortunately, a series of Clean Air Acts reduced air pollution for some decades.

Subsequently, in 2018, a coroner (Philip Barlow) found that the death in 2013 of a nine-year-old girl, Ella Adoo-Kissi-Debrah, who lived close to the South Circular Road in Lewisham (south London), was due to 'asthma, contributed to by exposure to excessive air pollution'. During the three years prior to her death, Ella had had multiple seizures and had been admitted to hospital 27 times, and in 2021 had been classified as disabled. The coroner said, in the course of his verdict, that levels of nitrogen dioxide near Ella's home 'exceeded World Health Organization and European Union guidelines', adding that there was 'a recognized failure to reduce the levels of nitrogen dioxide, which possibly contributed to her death', and also that another possible contribution was the 'lack of information given to Ella's mother'. The unlawful levels of pollution were detected at a monitoring station one mile from Ella's home (BBC 2020).

The coroner also advocated the adoption of legally binding targets for particulate matter, so that the United Kingdom can comply with World Health Organization guidelines. Significantly, he also recognized that there is 'no safe level of particulate matter' in the air. Roger Harrabin, the BBC environment analyst, commented that this was a historic verdict, pinning Ella's untimely death on the air that she breathed. On the sources of air pollution, he added that besides emissions from vehicles, sources include gas boilers, construction equipment, and paint and dust from brakes and tyres. Sadiq Khan, who as mayor of London was named as an interested party in the inquest, called the outcome 'a landmark moment', commenting that 'today must be a turning point so that other families do not have to suffer the same heartbreak as Ella's family' (BBC 2020).

An earlier inquest, held in 2014, had concluded that Ella's death was caused by 'acute respiratory failure and severe asthma', but in 2018 Judge Mark Lucraft, QC, ruled with two other judges that the 2014 conclusions should be quashed, and that a fresh inquest must be held (BBC 2019). Sarah Woolnough, chief executive of Asthma UK and the British Lung Foundation, called on the government, after the new inquest, to outline a public health plan to protect against 'toxic air' immediately. She saw fit to add that 'Today's verdict sets the precedent for a seismic shift in the pace and extent to which the government, local authorities and clinicians must now work together to tackle the country's air pollution crisis' (BBC 2020).

These tragic events bring out the role of both nitrogen dioxide and of particulate matter in urban air pollution; yet these phenomena are replicated in thousands of towns and cities worldwide. To focus on an individual case as an example of a global crisis may appear selective, granted the vast diversity of circumstances attendant on urban pollution across developed and developing countries. But it is individual cases of this kind that can galvanize national and sometimes international authorities to take action, and alert the public to the role of vehicle, industrial and household emissions in more than one global crisis. Air pollution may not involve tipping points of vast natural systems, but its global character shows how societies worldwide need to change direction while tipping points of a social character, like the generation of intolerable toxicity levels in cities across the globe, are at risk of being transgressed.

How the climate crisis, biodiversity loss and pollution comprise a single emergency

The World Health Organization estimates that there are 7 million premature deaths every year due to the combined effects of outdoor and household air pollution, with millions more falling ill from breathing polluted air. More than half of these deaths are recorded in developing countries (World Health Organization 2021a, 2021b). Thus the impact of air pollution bears comparison with that of climate change in terms of human casualties.

These crises are also related through having overlapping causes. Thus emissions from vehicles and from domestic and industrial fires (including fires from waste disposal) contribute centrally to both of them. This overlap considerably reduces the point of the kind of climate change denial that rejects the causal link between human activity and climate change, granted that the millions of deaths and illnesses resulting from air pollution are manifestly due to a large subset of what almost all scientists agree to be the causes of climate change. Indeed, it would be amazing if climate change deniers were to stage a campaign to preserve those emissions (methane emissions in particular) that are standardly blamed as among the causes of climate change but not normally blamed for air pollution.

So the crises of climate change and air pollution can readily be seen as comprising a single emergency, poised to get rapidly worse in the absence of strong concerted action. But should biodiversity loss and species extinctions be considered another aspect of the same problem?

To begin with, there are once again overlapping causes. Rising levels of greenhouse gases lead both to the melting of ice caps and glaciers and to stress on the creatures (such as polar bears) that depend on them. They also lead both to ocean acidification and to the bleaching of coral reefs, with losses to the biodiversity of oceans in general and coral reefs in particular. They further lead to forest dieback and to losses of the even vaster biodiversity of rainforests.

Certainly, some biodiversity losses are due more to unsustainable agriculture and unsustainable technology than to greenhouse gases. This is probably the case, for example, with the losses to the populations of pollinators such as monarch butterflies and honeybees, while other examples are due as much to unsustainable forestry (deforestation) as to changing rainfall patterns. But these losses resemble climate change in involving direct harm to human beings; and the unsustainable practices that underline them are all of a piece with the unsustainable forms of energy generation that underlie climate change.

With biodiversity loss, the losers clearly range more widely than the human victims of climate change since they include many non-human species. Yet there are plentiful non-human victims of climate change as well, such as those creatures whose habitats are eroded by droughts and wildfires or by storms and floods, and which have to move to new habitats further from the equator, when they do not run out of habitats altogether. Besides, the concerns of most contemporary people extend beyond the interests of human beings to those of (at least) sentient creatures, allowing the adverse impacts of both climate change and biodiversity loss to be recognized as disastrous where these creatures and their losses are concerned, as well as through the human harm and suffering involved.

Besides, the Natural Resources Defense Council cites a recent study suggesting that biodiversity and climate change are linked in quite surprising ways. When large mammals survive or are preserved in a landscape, it maintains, forests with wolves are more biodiverse and also store vastly more carbon than forests without them, while forests with tigers are more biodiverse and store carbon at a rate three times higher than forests that have lost their tigers. In South America, mammalian species such as tapirs are effective at reseeding rainforests, given their carbon-sequestering capacities. Meanwhile, ocean acidification is more effectively weathered and withstood where biodiversity remains robust. The re-introduction of large mammals to their former habitats could restore depleted biodiversity and at the same time replenish depleted carbon resources. This could be achieved through restoring afforested corridors between fragmented habitats, reinstating native vegetation and protecting wild mammalian species from hunting and other human pressures. These findings further strengthen the case for preserving in a wild state 30 per cent of the Earth's land surface, and 30 per cent of the oceans as marine sanctuaries (Bittel 2022; Vynne et al. 2022).

The three crises discussed in the previous chapter and in this one (climate change, biodiversity loss and air pollution) thus have strongly overlapping causes, including both greenhouse gases and unsustainable practices, and involve overlapping impacts, including widespread human and non-human deaths and illness. Further, their impacts also involve dislocations or even severings of the chain of nature, as natural systems approach their tipping points, and as vital links in that chain (think of pollinators; think too of pristine forests) come close to breakage.

Concerted actions need to take into account not only the mitigation of greenhouse gases and adaptation to their irreversible impacts, but also the preservation of natural systems before their tipping points are transgressed, and the amelioration of unsustainable systems of energy generation, agriculture, forestry and technology to make them into sustainable systems, so that all these crises are controlled and (if possible) curtailed. Besides all this, consideration needs to be given to rectification of the loss and damage undergone by developing countries, which have contributed little to causing the problems but suffered more than their share of the adverse impacts.

But these issues bring in issues of ethics and of equity, the themes of chapters 4 and 5. While a comprehensive plan of recovery and retrieval is beyond the scope of this book, some promising outlines can be delineated (see chapter 6), as long as our values and priorities are made central, and sustainable national and international policies can newly be based on them.

Recommended reading

Almond, R. E. A., Grooten, M. and Petersen T. (eds). 2020. *Living Planet Report – Bending the Curve of Biodiversity Loss.* Gland, Switzerland: WWF. https://www.wwf.org.uk/sites /default/files/2020-09/LPR20_Full_report.pdf

Barnosky, Anthony D. 2014. *Dodging Extinction: Power, Food, Money, and the Future of Life on Earth.* Oakland, CA: University of California Press.

BBC. 2021. News. 'COP26: What Was Agreed at the Glasgow Climate Conference?' BBC: Science & Environment, 15 November. https://www.bbc.co.uk/news/science-environment -56901261

Carson, Rachel. 2000 [1962]. *Silent Spring.* London: Penguin Classics.

Defra (UK). 2022. *UK Air: Effects of Air Pollution*, 22 August. https://uk-air.defra.gov.uk/air-pollution/effects

Harvey, Fiona. 2022. 'Major Cities Blighted by Nitrogen Dioxide Pollution, Research Finds'. *Guardian*, 17 August. https://www.theguardian.com/environment/2022/aug17/major-cities-blighted-by-nitrogen-dioxide-pollution-research-finds

Mathews, Freya. 2010. 'Planetary Collapse Disorder: The Honeybee as Portent of the Limits of the Ethical'. *Environmental Ethics* 32(4): 353–67.

Newman, J. A., Varner, G. and Linquist S. 2017. *Defending Biodiversity: Environmental Science and Ethics*. Cambridge: Cambridge University Press.

Read, Rupert. 2022. *Why Climate Breakdown Matters*. London: Bloomsbury Academic.

World Wide Fund for Nature (WWF). 2022. *What is the Sixth Mass Extinction and What Can We Do About It?* https://www.worldwildlife.org/stories/what-is-the-sixth-mass-extinction-and-what-can-we-do-about-it

Further reading and other sources

BBC. 2020. 'Ella Adoo-Kissi-Debrah: Air Pollution a Factor in Girl's Death, Inquest Finds', 16 December. https://www.bbc.co.uk/news/uk-england-london-55330945

Breaking Boundaries: The Science of Our Planet [film]. 2021. David Attenborough and Johan Rockström, dir. Jon Clay. All3Media.

Center for Biological Diversity. 2022. 'Monarchs on Global List of Threatened Species', 28 July. Tucson, Arizona: Center for Biological Diversity. bioactivst@biologicaldiversity.org

Fuller, Gary. 2022. 'Even Low Levels of Air Pollution Can Damage Health, Study Finds'. *Guardian*, 12 August. https://www.theguardian.com/environment/2022/aug/12/even-low-levels-of-air-pollution-can-damage-health-study-finds

International Union for Conservation of Nature (IUCN). 2022. Migratory Monarch Butterfly Now Endangered – IUCN Red List', 21 July. https://www.iucn.org/press-release/202207/migratory-monarch-butterfly-now-endangered-iucn-red-list

Koop, Fermin, Lam, Regina and Zhijian, Xia. 2022. 'COP15 reaches Historic Agreement to Protect Biodiversity'. 21 December. https://chinadialogue.net/en/nature/cop15-reaches-historic-agreement-to-protect-biodiversity/

Mulhern, Owen. 2020. *The Statistics of Biodiversity Loss [2020 WWF Report]*. https://earth.org/data_visualization/biodiversity-loss-in-numbers-the-2020-wwf-report/

Thomson, Ashley. 2020. *Biodiversity and the Amazon Rainforest*, 22 May. Washington, DC: Greenpeace USA. https://www.greenpeace.org/usa/biodiversity-and-the-amazon-rainforest/

UNEP (United Nations Environment Programme). 2021. *Why are Coral Reefs Dying?*, 12 November. https://www.unep.org/news-and-stories/story/why-are-coral-reefs-dying

Zissu, Alexandra. 2022. 'Colony Collapse Disorder: Why Are Bees Dying?', Natural Resources Defense Council, 29 April. https://www.nrdc.org/stories/buzz-about-colony-collapse-disorder

4

Ethics, Needs and Climate Justice

Ethics and the nature of ethical language

In the words of the *White Paper on the Ethical Dimensions of Climate Change*, ethics is 'the field of philosophical inquiry that examines concepts and their employment about what is right and wrong, obligatory and non-obligatory, and when responsibility should attach to human actions that cause harm' (Brown et al. 2005: 7). In the West, ethics has been studied since the days of Plato and Aristotle. Aristotle in particular discussed related concepts such as choice and the virtues, analysing them so as to present how the just person thinks and behaves (Aristotle 2000). He had no doubt that there are (at least often) correct answers to ethical questions, and that developing the virtues predisposes you to finding such answers.

Answers such as these (like the principles on which they depend) are normative; they concern what you or I *should* do, or what *ought* to be done, whether by individual citizens, by governments, or by other organizations. Unlike scientific findings, they are not empirical or observable facts, nor explanations of them. Some people suppose that this makes them nothing better than subjective claims; and certainly there is nothing to stop speakers making remarks about what you *should* do without having any good grounds. But what you should do, ethically speaking, is not groundless. Good grounds can include considerations such as that an action would harm someone, or is part of a societal practice that is beneficial overall (like keeping promises). And because there can be ethical grounds such as these, ethical judgements can be more or less defensible, and more or less reasonable.

There are, certainly, other kinds of *shoulds* and *oughts* than ethical ones. Some of them are legal, some are aesthetic, and some are prudential (concerning what you should do in your own interests). The criteria for legal rightness, aesthetic rightness and prudential rightness differ, and support different conclusions (some better and some worse) about what should be done legally, aesthetically or prudentially. Because the focus here is on ethical *oughts* and moral rightness, we can largely set aside other kinds of *shoulds* and *oughts*, and concentrate on ethical ones.

Before we do so, however, it is worth mentioning that our ability to reflect on *shoulds* and *oughts*, recognizing that they are not claims of empirical fact, corresponds to a valuable aspect of human life, that of our agency and our freedom. We are free to envisage different futures, and to recommend and advocate some over others on the basis of relevant grounds and criteria. This ability enables us not only to discover what is (already) real but to decide and to change what is to be real, and to make it real. And to do this, we need not only talk of what was or is the case (and thus the vocabulary of science and or history), but also the language of *shoulds* and *oughts*, ethical *shoulds* and *oughts* included. Like other normative language, ethical language is thus essential to our humanity.

It is also vital to our present predicament because we need to discover what we (as individuals, as governments, and through other organizations) *should* do about the climate crisis. Our freedom is far from unlimited, but we can use such agency as we have to enhance the world, or at least to reduce the prospect of climate change or anything else destroying its ecosystems and its biosphere, on which we all depend.

Moral standing and reasons for action

In order to reflect on moral rightness in the field of climate justice, we first need a view of which beings or entities count, morally speaking, or should be taken into account when decisions are being made. Such a view can help us with regard to the scope of ethics, and also with regard to what counts as a reason for right action.

The most insightful essay on what counts, morally speaking, is an article by Kenneth Goodpaster (Goodpaster 1978). Goodpaster entitles this essay 'On Being Morally Considerable', meaning by 'considerable' not 'large', but 'needing to be taken into account'. However, as 'morally considerable' (and 'moral considerability') are confusing phrases, I will write here of 'moral standing' instead, to express exactly the same concept. Another expression that means much the same is 'moral patients' or beings on the receiving end of moral action, a phrase used slightly earlier by the late Sir Geoffrey Warnock (Warnock 1971). So you can think of bearers of moral standing as moral patients if you like.

Goodpaster's central theme is that the bearers of moral standing are beings or entities with a good of their own. He does not include among such beings rocks or rivers because they do not have a good of their own and cannot be benefited, but he does include all living creatures because they can (Goodpaster 1980). There is such a central connection between morality and benefiting that all beneficiaries (potential or actual) should be understood to have moral standing (or, in his terms, 'considerability') (Goodpaster 1978). Potential beneficiaries include, of course, future beings, and so the bearers of moral standing include, for Goodpaster, living beings both present and future. Goodpaster proceeds to consider a wide range of imaginable objections to these conclusions, and then supplies convincing replies to all of them (ibid.).

This persuasive account of the scope of moral standing leads to a matching account of reasons for moral action. For it is the welfare or well-being of living beings that qualifies them for moral standing, and so it will be their well-being that has independent value (also known as 'intrinsic value') and that supplies reasons for action. It cannot supply reasons for all agents on all occasions, but on those occasions when particular agents can promote or protect this well-being, or prevent its deterioration, it comprises a reason for them to do so.

There will also be occasions when certain actions can supply a means of or a condition for the well-being of a being or creature with moral standing, and on these occasions the means or condition will have derivative value (in these cases, instrumental value), and once again supply a reason for action. Thus there is a close link between reasons for action and value, whether intrinsic, instrumental, or of some other kind (such as aesthetic value); where there is value (of one kind or another), there will also be reasons for action (for one agent or another, or for more than one).

Yet very often the difference that can be made to the well-being of a being or creature will be slight, and so the reasons may easily be outweighed by reasons to do something else. Reasons for action become much stronger when they relate to the *needs* of the being or creature. Needs are either necessary conditions for well-being to remain intact, rather than being undermined, or indispensable components of well-being itself. In both cases, the reasons for action to which needs give rise will be strong ones. They may still be overridden, but not so easily. (Many people, including advertisers, speak of objects of strong desire as 'needs', misrepresenting wants as needs; but these are not what is meant by 'needs' in the present context.)

Sometimes the balance of reasons for an action is so strong that it would be wrong not to perform that action. Actions of this kind are obligatory, or *obligations*. So there is a strong connection between values, or reasons for action, and obligations, and at least as strong a link between needs and obligations. And when we recognize the strong ties between obligations and such reasons for action as needs, any inclination that we may have had to think of obligations as subjective falls away.

The climate crisis places the needs of millions of fellow humans in jeopardy, and this already supplies the strongest of reasons to counter it and to mitigate climate change, often amounting to an obligation. It is not too late for governments and for individuals and companies to make a difference and reduce the dangers, even though the window of opportunity for doing so is reducing (Roser and Seidel 2017: 25).

Yet it is not only fellow humans of the present whose needs are at stake. So are those of the generations of the foreseeable future. In the following section, I argue that the interests of future generations should be included within the scope of justice. (Readers inclined to accept this conclusion and disinclined to follow some detailed philosophical argument could skip the next section and move on to the following one.)

Future generations, obligations and justice

Justice is centrally concerned with honouring rights, and observing the rights of fellow humans is certainly central to justice. However, we have already seen that the scope of moral standing extends further than that of the bearers of rights. As we have seen, moral standing belongs to future generations as well as to those currently alive, and thus their well-being can form reasons for action in the present, amounting sometimes to obligations; indeed, disregarding the well-being and the needs of future people is widely held to amount to treating them unjustly. And yet there is a case for holding that most future people do not have rights in the present. They will have rights against the contemporaries of their lifetimes, but not beforehand.

Rights can only be held by individuals or groups that can be treated better or worse, and thus by those about whom it makes sense to speak of rights being respected or disrespected. But most future people cannot, as individuals, be treated better or worse by people now alive because their very identity is not yet determined, and whether one set of individuals or a different set comes into being depends itself on choices of the present and the near future. Different policies chosen in the present and the near future make a difference to which people meet and mate, and also to when they do so. There again, the identity of those future people who have not yet been conceived will depend partly on who their parents are and partly on the timing of their conception. So whether one set of future people or another set comes into being depends on all these choices (McKinnon 2022: 169–70; Parfit 1984). Actions of the present can certainly make a difference to the likely well-being of whichever people come into being in the future, but not to whether particular future individuals are treated better or worse. Accordingly, those future people who have not yet been conceived cannot be said to have rights in the present (Schwartz 1978, 1979).

Thomas Schwartz, who drew attention to this problem, concluded that current people have no obligations to the majority of future people at all. However, Derek Parfit, who called this problem 'the non-identity problem', reached a different conclusion. He granted that we cannot harm (most) future people as individuals but insisted that we can still make a difference (often a large difference) to their quality of life, and thus to their well-being. Through current policies and decisions, we can affect which people come into being, and also into what kind of society and what kind of environment they are born and grow up. So the difference that we can make to the quality of life of whoever will live in (say) the later decades of the present century is at least as great as the difference that we can make to our contemporaries. His conclusion was that, while we do not have obligations to (particular) future people (or rather, to the unconceived ones), we have considerable obligations in their regard (Parfit 1984). In other words, it would be wrong not to take their interests into account. Most ethicists have accepted this view, at least implicitly.

But are these obligations of present people with regard to future people obligations of justice? John Broome concludes that they are not because (most) future people do not have rights. And this, he believes, has an important bearing on the obligations of current people because obligations of justice outweigh other obligations (Broome 2012). Admittedly, he also holds that governments have obligations to maximize the balance of well-being over suffering across the foreseeable future, and future people figure importantly in this reckoning. But he considers that individual agents should prioritize justice, and thus avoid doing harm to people with rights; and these do not include future people. (He seems not to make it clear whether the obligations of collectives such as companies are more like those of governments or more like those of individuals.)

But can the scope of justice be confined to the bearers of rights? Derek Parfit wrote as if we should treat the interests of present people on a par with those of (foreseeable) future people, where the quality of life of the latter is at stake. Thus we should not use up key resources in the present and the near future that would deplete the quality of life for many centuries of prospective future people because a greater adverse difference would be made to whoever lives in the future than to the people of the present and the near future. No particular person living in the further future would be adversely affected, but Parfit makes it clear that this is inconclusive because the overall future quality of life would be depleted, and that we should not embark on 'depletion' (his name both for this thought-experiment and for the practice at its core) (Parfit 1984: 361–4). Parfit's argument here seems convincing. And although he does not expressly mention justice here, he does mention it with regard to impacts of current action on future people in the next section (ibid.: 365), a section in which he also concludes that appealing to rights cannot resolve problems such as that of 'depletion' (ibid.: 364–6).

A further argument for including future generations within the scope of justice has recently been presented by Blake Francis. We can imagine future people, if they lack the viable climate that most contemporary humans experience, complaining that their predecessors deprived them of such an environment. This would amount to a complaint of unfairness or injustice, and reasonably so. Hence conventional accounts of justice that have no room for future people need to be revised (Francis 2020: 17–19).

This being so, we may reasonably conclude that justice is not limited to the bearers of rights, and that the obligations that Parfit maintains that we have with regard to future generations have no less significance, even if these generations lack present rights, than those owed to the current bearers of rights. I have argued for this conclusion in my recent book *Applied Ethics* (Attfield 2022b: 36–7). Also, in two articles jointly authored with Rebekah Humphreys, I have argued that the same is true of non-human animals, and that even if we do not hold that animals bear rights, their needs, which are often comparable to those of human beings, mean that they too come within the scope of justice (Attfield and Humphreys 2016, 2017). However, I will return to the bearing of obligations to non-human animals in the next chapter.

It is also significant that the many people who talk and write about climate justice do not appear to confine justice to bearers of rights. They are certainly concerned about preventing harm to the contemporary victims of climate change, but they are equally concerned about the prospective victims of the future, without deviating from concern that climate justice should be respected and practised. (See, for example, Cripps 2022: 38–9; McKinnon 2022: 53–61.) Accordingly, when we reflect on what both morality and justice require against the background of the climate crisis, we should take into account the needs and interests of future generations as well as those of the people currently alive.

Further issues about future people

However, there are further issues concerning the significance of the needs and interests of future people in ethical decision making. Many people, and many economists in particular, take the view that the uncertainty of future needs and interests reduces present responsibilities, or allows them to be overridden when they clash with current interests. Some add that current responsibilities are reduced in proportion to the distance in time of future interests from the present. Often, this is treated as a reason for discounting future costs and benefits by an annual percentage, sometimes as high as 5 per cent.

Other grounds given for discounting include opportunity costs and time preference. If, for the sake of retaining resources for future years, we defer investing in the present in ways that would have immediate benefits, then we forego important opportunities. Yet many forms of present investment (for example, in infrastructure) would benefit both the present and the future, and so the possible conflict between present and future interests should not be exaggerated. It should be added that such investments in infrastructure could include measures of adaptation to global heating, designed to protect both our own and future generations from the worst effects of flooding, storm damage, heatwaves or wildfires. Further, as Parfit's thought-experiment about depletion shows, it is sometimes unjustifiable and wrong to consume resources in the present which could provide for the needs of many future generations (Parfit 1984: 361–4). Thus the argument from opportunity costs is far from conclusive, and in some circumstances carries negligible weight.

There again, the fact that people's behaviour often displays time preference is no good reason for downplaying future needs and interests. Injury, illness, mutilation and death will be just as bad in fifty or a hundred years' time as in the present, and investment in steps taken to prevent them can rightly take precedence over efforts to prevent less serious suffering in this year or next.

Further, considerable doubt has been cast on whether time preference is really as general or as deeply entrenched as it used to seem. (If things were otherwise, some kind of 'democratic' argument could be mounted that we ought to mould public policy in line with these preferences.) Yet the research of Hilary Graham suggests that surveys of attitudes to future costs and benefits generate different outcomes, depending on how the questions are worded and framed. While questions with detached and impersonal wording are prone to produce the customary results, quite different findings emerge when phrases are used such as 'your grandchildren's generation', with many respondents wanting the interests of that generation to receive equal consideration to those of the current generation, and some favouring them being prioritized over present interests (Graham et al. 2017).

Besides, the view that people's good is a function of their preferences should itself be contested. Human needs, when this expression is not used in the distorted sense favoured by advertisers, seldom turn on preferences; indeed, needs and preferences frequently conflict. (See the second section of this chapter at pp. 54–6 for a more detailed account of human needs.) Thus the need for a habitable environment remains a need even when people's preferences are for kinds of consumption, travel and transport liable to undermine this need. (I have argued elsewhere for the independence of needs from preferences: Attfield 1995, 2022a.) So the argument from time preference carries little or no weight at all.

That leaves the argument from the uncertainty of future needs and interests. Yet uncertainty does not increase in proportion to temporal distance from the present. Some future impacts of present actions are readily foreseeable (not least the case of contributions to global heating). There again, outcomes located in the present or the near future are often uncertain (like tomorrow's weather, weather forecasts notwithstanding). Hence, even if it is rational to discount impacts because they are uncertain, that is not a ground for discounting future costs and benefits in general (or blanket-wise), and certainly not for discounting them in proportion to temporal distance from the present.

It is worth bearing in mind that discounting future costs and benefits at a compound rate of anywhere near 5 per cent makes the costs and benefits of thirty years hence count for very little, and those of one hundred years hence count for virtually nothing. Yet present actions (and omissions) can make a huge difference to future generations for many decades, not to mention centuries (Hourdequin 2007). For example, as we saw in chapter 2, they could contribute to whole ecological systems reaching their tipping points, with dire consequences for humanity ever after. And this consideration supplies a strong argument against (at any rate) high discount rates, which, unsurprisingly, fail to support future-oriented policies, and leave a question mark even over lower ones.

Here it is worth citing a passage by Broome.

It is largely because of their different discount rates that the *Stern Review* and Nordhaus's *A Question of Balance* lead to such very different recommendations for economic policy. Because its discount rate is low, the *Stern Review* asks the present generation to make urgent sacrifices for the sake of future people. Because he uses a much higher rate, Nordhaus does not. (Broome 2012: 105)

Broome's sympathies are clearly not with William Nordhaus (Nordhaus 2008), but rather with the *Stern Review* (Stern 2007), even though that work has come in for widespread criticism from many economists and environmentalists. In any case, the passage bears out the importance of discount rates, and, in the light of the above discussion, the rightness of selecting either a low rate or none at all. It is only either low discount rates or their complete absence that allows the needs and interests of future generations to be taken adequately into consideration, and thus to be treated justly.

Must present and future interests conflict?

There is often a case for deploying resources on present needs and interests, rather than future ones. Thus it is held that the present should not be sacrificed, Robespierre-like, for the good of an idealized future that may not materialize. Besides, the present can be seen as the final opportunity to make provision for current needs; and satisfying these needs is often a necessary condition for the individuals in question to have a future at all. There again, the responsibilities that we have through our existing relationships can be adduced in support of giving priority to those present people with whom we have these relationships. Writers sceptical of the proposals of environmentalists have accordingly suggested that attention and resources should be focused on the alleviation of current poverty rather than on projects to enhance quality of life over the long term (Lomborg 2001).

Yet the argument from relationships cuts both ways; for most adults have relationships with children (whether theirs or other people's), who are likely to outlive them, and many with grandchildren whose lives may continue into the coming century. As the passage in the last section about Hilary Graham's research suggests, many people are already aware of the importance of fostering the prospects not only of their own grandchildren but also of their grandchildren's generation. Besides, if we neglect the prospects of a habitable environment for the people of future decades and centuries, then our children and grandchildren will have ample grounds to deplore our concentration on current interests to the exclusion of theirs. Thus if we fail to mitigate our greenhouse gas emissions, through prioritizing our own aspirations for travel, tourism and current comfort, we stand to be criticized for ignoring a window of opportunity, possibly no longer open to them, to protect them from severe weather events such as coastal flooding, prolonged heatwaves, droughts and hurricanes. Lomborg's concern to alleviate current poverty does him credit, but if at the same time we neglect the global environment, and thus the environment of poor countries, the impacts of poverty alleviation in the present may fail to be long-lasting.

In any case, while responsibilities to the present and to the future can conflict, there is no inevitability that they will or must. We have already seen, in connection with investments in infrastructure, that it is possible to benefit both the present and the future through the same action. Thus the British-built railways of India have benefited both the India of the colonial period and also the India of the post-colonial epoch.

There again, there is a whole class of practices capable of benefiting both the present and the future, practices that are sustainable. Examples relevant to averting the climate crisis include the generation of electricity by sustainable means; thus renewable forms of energy generation, such as solar power, and power from hydroelectric, wind, wave, tide and geothermal sources, are capable of supplying energy for generations to come, once suitable equipment has been installed, for as long as the sun continues to shine or the wind to blow. For a practice to be sustainable, it must not undermine potentially sustainable practices elsewhere, nor its own continuation into the future; however, these requirements are readily satisfied by the renewable generation of electricity.

A further example of a sustainable practice is sustainable forestry. As long as each generation leaves enough forest to allow the forest to regenerate, every generation can benefit from its products, as long as severe weather events allow the forest to remain intact (Williams 1978). Such policies need to be modified so as to prevent forest species being extinguished; yet it remains significant that some harvesting of the forest for the sake of the current generation need not undermine the interests of future generations. Thus the introduction of sustainable practices holds out a promise of ways in which societies can adapt so as to live in harmony with themselves and their environments; each community should, where possible, introduce systems of sustainable agriculture, sustainable education and sustainable production.

The more sustainable practices are introduced, the clearer it will be that present and future interests need not conflict. Far from harming the people of the present, measures designed to uphold the quality of life of future people can often be beneficial in the present as well.

At the same time, it should be emphasized that policies of averting environmental catastrophes should not be regarded as sacrifices on the part of the current generation for the sake of future generations. That is because current interests are also crucially at stake, with current floods, heatwaves, wildfires, droughts and hurricanes blighting the lives of many of our contemporaries. Some of these events are situated in developing countries (where the needs of many are failing to be met), which is a reason of itself to take action; but many are also affecting Europe, North America and Australasia, and thus the places where most of the readers of this book are likely to live.

The way in which climate change has become an urgent matter has not gone unnoticed. Nicholas Pidgeon has disclosed research showing that a high proportion of British adults no longer regard climate change as a concern mainly about the prospects of future people, and have come to regard it in the last two or three years as a pressing issue for the well-being of themselves, their families and their contemporaries (Pidgeon 2022). The imminence of several tipping points, the biodiversity crisis and the ongoing threats to human health from air pollution underline the sound basis for these changed perceptions.

The Precautionary Principle

The importance of taking tangible environmental threats seriously was recognized first in German and then in European Union legislation, and subsequently at the Rio Conference on Environment and Development held at Rio de Janeiro in 1992, in the form of the Precautionary Principle. This principle can take a variety of forms, but the core notion is that where there is good reason to believe that severe and/or irreversible impacts are possible, preventative action should be taken in advance of scientific consensus: see the Rio Declaration, which included a restrictive version of this principle (Granberg-Michaelson 1992; United Nations 1992). UNESCO, in its *Declaration of Ethical Principles in Relation to Climate Change*, has more recently adopted a similar but adjusted version of this approach in Article 3: 'Where there are threats of serious or irreversible harm, a lack of full scientific certainty should not be used as a reason for postponing cost-effective measures to anticipate, prevent or minimize the causes of climate change and mitigate its adverse effects' (UNESCO 2017).

Sometimes, caveats are added. It is sometimes added that preventative action must employ the best available technology (BAT). But the pace of specialist technological advance often means that the best available technology is unaffordable, while other forms of technology would be almost as effective and should be used instead. Another caveat is that the preventative action should not entail excessive costs (NEEC), echoed in the UNESCO version through the insertion of the phrase 'cost-effective measures', and this appears reasonable. Yet the significance of the threats to be avoided makes the question of what expenditure would be excessive a controversial one, since considerable expenditure might well be justified to prevent a succession of systems reaching their tipping points and undermining the prospects of most life on our planet, even if, in terms of immediate outcomes, it was less than cost-effective. In the interests of avoiding interminable debate about what expenditure would be excessive or cost-effective, it is advisable not to include caveats such as these in the first place.

Stephen Gardiner considers one standard statement of the Precautionary Principle: 'When an activity raises threats of harm to human health or the environment, precautionary measures should be taken even if some cause and effect relationships are not fully established scientifically' (Gardiner 2011: 412; Wingspread Statement 1998). However, this formulation is open to a denial from sceptics that the proposed activity (fracking, say) does really raise such threats, despite evidence that it may well contribute to them. It is also open to the rejoinder, as Gardiner notes, that it could 'be invoked to stop any activity, however beneficial, on the basis of any kind of worry, however fanciful' (ibid.). This objection makes the principle seem irrational.

Yet, in other forms, the Precautionary Principle is not open to these objections. It can be worded so as to open, as above, with the phrase 'Where there is good reason to believe that severe and/or irreversible impacts are a possibility (from an action or policy)' and can immediately proceed with 'preventative action should be taken in advance of scientific consensus about the severe and/or irreversible nature of these impacts'. In this version, fanciful worries are precluded because good reasons are required, and for the same reason most beneficial actions are exempted because most actions will not be accompanied by there being good reason to believe that impacts of severe and/or irreversible kinds are on the cards. In this form, it also becomes clearer that the cause-and-effect relationships in question must be ones relevant to the undesirable impacts of the proposed activity. At the same time, it becomes harder for sceptics to deny the relevance of the principle, because, even if they deny that severe and/or irreversible impacts are actually at stake, they can hardly rationally deny that there is good reason to believe that human-generated climate change is taking place, causing increased risks of sea-level rise and of extreme weather events. Even if they choose to reject the findings of IPCC science (see chapter 2), it is hard to deny that there are good reasons to credit these findings.

The Precautionary Principle is itself an *ethical* principle, claiming that in certain circumstances preventative action *should* be taken, and that it should be taken not on technical, aesthetic or purely prudential grounds, but on grounds relating to the well-being of human beings and other creatures that have *moral standing*. It can also be recognized as a principle of *justice*, even on the narrower view that relates justice to avoiding harm to the bearers of rights (because it undeniably seeks to do this), as well as on the broader view (defended earlier in this chapter) that relates justice to upholding the quality of life of future generations, in addition to that of our contemporaries. If, like me, you are willing to include issues of the well-being of non-human animals within the scope of justice, then this stance too makes it all the more a principle of justice because it urges measures being taken to preserve the environments of wild, farmed and domestic animals.

There is no need to reach agreement on fundamental questions of ethical theory in order to recognize the Precautionary Principle (in the form expressed in the last paragraph but one) as a sound ethical principle. For it is readily supportable by most ethical theories. Thus theories that regard actions or practices as right because they produce the best balance of good over bad foreseeable consequences, or as good a balance as alternatives (consequentialist theories), clearly support it. But so do Kantian theories, such as that of Onora O'Neill, who stresses the duty to avoid and avert harm, just as the Precautionary Principle does (O'Neill 1986). So do theories based on rights, like that of the climate ethicist Simon Caney (Caney 2010). It could also be upheld by Rawlsian contract theorists, who hold that what would be agreed by rational parties in a fair bargaining situation is right and just (Rawls 1972). My own stance is a consequentialist one, but there is no need to persuade you of it in order to vindicate the Precautionary Principle.

John Broome, however, recommends that we adhere to policies of maximizing 'expected value', and not to adopt *any* version of the Precautionary Principle (Broome 2012: 129). While he is able to show that some versions of the Precautionary Principle are defective (ibid.: 119), and that it is widely better to reason from probabilities and recognized values, his own attempts to carry through this project with respect to climate issues of the foreseeable future appear to involve indeterminacy, not least when this method is applied to the obligations of governments. Since, however, we are faced with a pressing climate crisis, and need clear guidance about how governments, corporations and individual citizens should act, it is far preferable to follow defensible forms of the Precautionary Principle, like the one presented earlier in this section, confident that forms such as this one almost certainly cohere well with consequentialist approaches, like that of maximizing expected value.

Faced, as we are, with the prospect of tipping points being reached within fifty years, and triggering the tipping points of a potential cascade of global systems, the Precautionary Principle is just what is needed. It tells us to take action to prevent tipping points being reached, action both at government level and at the level of corporations, non-governmental organizations (NGOs) and individual citizens and families. Zero-carbon policies need to be adopted and implemented as soon as possible, and well before fifty years have elapsed, together with their wide implications across most of the world's societies. We do not need to await multiple authorities calculating and comparing the expected value of carbon-based and carbon-free policies across this century and the next, and then trying to find common ground between the policies favoured by their (potentially differing) estimates. The combination of IPCC science and the Precautionary Principle already supplies us with the ethical guidance that we need.

I am not suggesting that the Precautionary Principle supplies all the answers to all environment-related policy matters. For example, there is a strong case for educational policies to include provision for environmental education for all, both in the present and in all foreseeable decades. Yet such a policy would not be a precautionary policy, but one grounded in the needs of current students, those of their society, and those of future generations (both human and non-human). So it can be supported by direct appeal to the well-being of current and future generations alike, with no need to appeal to derivative principles such as the Precautionary Principle.

More generally, many ethical issues, including issues of justice, concern circumstances where we are not principally facing a prospect of serious or irreversible impacts expected to arise from current policies. For example, rectifying the injustice by which many developing countries are already suffering greatly from climate change to which they have contributed little or nothing (Brown et al. 2005: 12) is plainly an obligation of justice, but taking action to reduce global heating for their sakes and to facilitate improvements to their infrastructures is not centrally a matter of precaution, even though what needs to be done to rectify this injustice coheres well with what the Precautionary Principle in any case enjoins. So there remains a strong case for appealing, in cases where the Precautionary Principle is not the most directly applicable principle, either to the balance of expected value (as commended by Broome) or to principles of just distribution which can be argued to cohere with such a balance (see Attfield 2019 [1995]: 133–48).

The example of offsetting

While questions of the obligations of individual citizens and households will be addressed in later chapters, it may be worthwhile to look at one recommendation for action at the individual level. Broome, who considers that the obligations of individuals are primarily to avoid harming contemporary bearers of rights, and also to avoid reducing the well-being of those future people whose identity is foreseeable, recommends that individuals of the present who undertake travel by air should make good the resulting carbon emissions by offsetting them (Broome 2012: 84–95). There are organizations that will plant trees for people who pay for this to be done, and advise about the amounts required in proportion to the length of the flights undertaken. In this way, they offset the resulting harm to the environment by ensuring that enough carbon-absorbing trees come into being and that over the course of time the photosynthesis of these trees will absorb an equivalent amount of carbon dioxide. The location of the trees is regarded as a matter of indifference; what matters (it is assumed) is the responsibility of current individuals not to harm others (present or future) by increasing the total of greenhouse gases.

While the Precautionary Principle relates most directly to states and large companies, it could reasonably be held to apply also to individuals contemplating air travel. For such travel will predictably increase levels of greenhouse gases and thus global heating, and will also contribute to air pollution and thus harm to people's health. These changes can have severe impacts and can contribute to irreversible ones; and so the principle enjoins precautionary action of the offsetting kind. Broome considers the objection that the practice of offsetting diverts attention away from the responsibilities of governments (Broome 2012: 94). But to this his answer is that it need not. It could, on the contrary, show that popular concern about global heating warrants action at government level as well.

However, even when the practice of offsetting means that more trees are planted, or that some trees that would have been cut down survive and continue to photosynthesize for longer, it takes no account of where the trees are located, or of the impact of offsetting on vulnerable systems that are approaching tipping points. Yet some systems, such as the boreal areas of north-west Canada, are being deforested (see chapter 2), and the planting of trees of native kinds could help to prevent the system that they comprise from reaching a tipping point. There again, the forests of Borneo are being cut down for the sake of palm-oil plantations, to the detriment of species living in those forests. At the same time, the system of the forests of Borneo is under threat and could reach a tipping point; and if this point is reached, this could drive other systems to reach tipping points of their own.

Accordingly, a greater positive difference would be made if offsetting took the form of tree planting or tree preservation in the boreal forests of north-west Canada or in the rainforests of Borneo, rather than in many other areas. So it is not indifferent where the trees planted or preserved through offsetting are located. And this has a bearing on the form that offsetting should take. There is, for example, a charity based in Cardiff called 'Regrow Borneo', which plants trees to prevent the fragmentation of forests in north Borneo. Contributions to charities of this kind stand to prevent harm to a greater degree than contributions to offsetting charities in general. Contributions to charities pledged to restore the boreal forest of north-west Canada would make a comparable difference. Ethical decisions should be informed by the risks and the harms threatened by current trends, and not just by the general problem of global heating.

Accordingly, the issues surrounding offsetting should not be seen as focused primarily on different understandings of justice or whether future generations of uncertain identity should be taken into account. They focus instead on the science of tipping points, and on actions that could postpone or prevent these tipping points being reached for the sake of affected parties, both present and future. At the same time, everyone should be aware of the limits of the potential effectiveness of offsetting. As Kathryn Brown (Director of Climate Change and Evidence for the Wildlife Trusts) explained in a recent presentation, there is simply not enough land available for offsetting through tree planting to make a difference sufficient to mitigate global climate change (Brown 2023). Reafforestation by governments, as in Ethiopia, Cuba and Haiti, is a different matter (Attfield 2015 [1999]: 90–1), but neither charities nor most individuals have access to the lands that need to be replanted in those countries, nor the capacity to influence the policies of those governments.

More generally, both ethics and science are needed to inform what we should do in face of climate change. In the next chapter, we consider the situation and needs of different sets of victims of climate change, and the shape that climate justice could take, in the light of the scientific and ethical background discussed in this and previous chapters.

Recommended reading

Attfield, Robin. 2015. *The Ethics of the Global Environment*, 2nd edn. Edinburgh: Edinburgh University Press.

Attfield, Robin. 2019 [1995]. *Value, Obligation and Meta-ethics*. Leiden, NL: Brill.

Attfield, Robin. 2022. *Applied Ethics: An Introduction*. Cambridge: Polity.

Brown, Donald A. et al. 2005. *White Paper on the Ethical Dimensions of Climate Change*. Rock Ethics Institute, Pennsylvania State University.

Cripps, Elizabeth. 2022. *What Climate Justice Means and Why We Should Care*. London: Bloomsbury Continuum.

Graham, Hilary, Bland, J. Martin, Cookson, Richard, Kanaan, Mona and White, Piran C. L. 2017. 'Do People Favour Policies that Protect Future Generations? Evidence from

a British Survey of Adults'. *Journal of Social Policy* 46(3)
(July): 499–520.

McKinnon, Catriona. 2022. *Climate Change and Political Theory*. Cambridge and Hoboken, NJ: Polity.

Parfit, Derek. 1984. *Reasons and Persons*. Oxford: Clarendon Press.

Roser, Dominic and Seidel, Christian. 2017. *Climate Justice: An Introduction*. Abingdon and New York: Routledge.

UNESCO. 2017. *Declaration of Ethical Principles in Relation to Climate Change*. Paris: UNESCO.

Further reading

Aristotle. 2000. *Nicomachean Ethics*, trans. and ed. Roger Crisp. Cambridge and New York: Cambridge University Press.

Attfield, Robin and Humphreys, Rebekah. 2016. 'Justice and Non-human Animals, Part I'. *Bangladesh Journal of Bioethics* 7(3): 1–11.

Caney, Simon. 2010. 'Climate Change, Human Rights and Moral Thresholds', in Stephen M. Gardiner, Simon Caney, Dale Jamieson and Henry Shue (eds), *Climate Ethics*. Oxford and New York: Oxford University Press, 163–77.

Francis, Blake. 2020.'Climate Change Injustice'. *Environmental Ethics* 44(1) (Spring): 5–24.

Gardiner, Stephen M. 2012. *A Perfect Moral Storm: The Ethical Tragedy of Climate Change*. Oxford: Oxford University Press.

Goodpaster, Kenneth E. 1978. 'On Being Morally Considerable'. *Journal of Philosophy* 75, 308–25.

Hourdequin, Marion. 2007. 'Doing, Allowing and Precaution'. *Environmental Ethics* 29(4): 339–58.

O'Neill, Onora. 1986. *Faces of Hunger: An Essay on Poverty, Hunger and Development*. London: Allen & Unwin.

Stern, Nicholas et al. 2007. *The Economics of Climate Change: The Stern Review*. Cambridge: Cambridge University Press.

United Nations. 1992. *The Rio Declaration on Environment and Development*. UN Document A/CONF.151/26. New York: United Nations.

5

The Victims of Climate Injustice and the Shape of Climate Justice

This chapter concerns different victims of climate injustice, ranging from vulnerable people to vulnerable non-human beings, and also the various forms and policies that climate justice might involve. The sections about victims serve two purposes. They underline the case for the ethical obligations in climate matters of the peoples and governments of the world, and particularly of the developed world; and they also raise questions of how particular sets of victims can be assisted, saved from disasters and, where possible, contribute to their own recovery from victimhood. The sections about responsibility for climate justice and the forms that it might take help explain what climate justice might look like in practice and, accordingly, the forms that our climate responsibilities may take.

Human victims of climate injustice: developing countries

One of the largest sets of victims of climate injustice consists in future generations, whose prospective situation and need for justice in the present has already been presented in chapter 4. As has also been mentioned in that chapter, the inhabitants of developing countries are often among the victims of global heating and related

crises too, despite having contributed only small proportions of greenhouse gas emissions (Brown et al. 2005: 12).
The total greenhouse gas emissions of some of the more
heavily populated developing countries bear comparison
with those of United States and Japan, but the per capita
emissions of countries such as India, Brazil and Indonesia
have been relatively small, while those of Ghana and of Fiji
are smaller still (Cripps 2022: 51). Yet it is the developing
countries of Africa, Asia, Latin America and Oceania
that are suffering most, whether from famines, floods,
droughts, wildfires or hurricanes. Some regions have even
undergone both flooding and droughts (at different times),
such as the area near Lake Turkana in the north of Kenya
(BBC2 2022c).

Certainly, framing developing countries and their
peoples as victims could be misleading. So it should
at once be made clear that many of their citizens are
actively promoting solutions to their various agricultural,
educational, social and economic problems, often with
considerable success, as charities such as Farm Africa,
Practical Action and Médecins sans Frontières frequently
report. The problem is that the greenhouse gas emissions,
largely emitted by people in distant continents, are stacking
the odds against the people of these lands and adding to
problems that in many cases existed already, poverty and
malnutrition among them.

While it is tempting to simplify matters and suggest
solutions about permissible emissions or targets based
on the different emissions of different countries in recent
decades, it is important not to disregard the earlier history
of the developing countries as well. As Simon Caney has
emphasized, the problems stemming from climate injustice
are intertwined with those resulting from colonial rule
(in which economic exploitation played a leading part),
from the widespread poverty that these countries inherited
when they became independent, from the skewed nature
of the international trading system, in which the producers
of primary products receive limited returns compared with

those who process them, and from the tax avoidance of multinational companies active in their territories (Caney 2012, 2020; McKinnon 2022: 37). Even countries such as Thailand and Ethiopia, which largely avoided colonial rule, suffer from many of these same problems.

This does not mean that agreements about climate mitigation and adaptation, urgently needed as they are, must resolve all these issues at the same time (on an all-or-nothing basis). But it does mean that the vulnerabilities of developing countries should not be forgotten when (say) policies for energy provision and food security are being considered. There again, policies of adaptation to levels of global heating that can no longer be reversed will be different for countries with serviceable infrastructures, and countries like the Democratic Republic of Congo and the ex-British territory of Guyana, which have few roads between cities at all.

We should next review why countries of the 'global South' are generally worse affected by climate change than those of the 'global North'. The countries of most of Africa and of South Asia, already hot and humid, are now subject to prolonged periods of heatwaves combined with increased humidity. Food security, already a problem in these same countries, is increasingly at risk, both through storms and floods and equally through droughts, like the one afflicting the Horn of Africa (in the autumn of 2022 and early 2023). Further, higher temperatures have contributed to the transmission of diseases such as dengue fever, Zika virus and West Nile fever, particularly where comprehensive health systems capable of curtailing the spread of these diseases are lacking (McKinnon 2022: 38–9).

There again, the social systems of countries such as Indonesia, Uganda and Mali mean that the women of those countries suffer more from malnutrition than their menfolk (Cripps 2022: 65). Thus climate injustice can blend both with racial injustice and with gender injustice. Some women of Africa and South Asia have become leaders in campaigning for agricultural improvements and in resisting deforestation (ibid.: 66).

One way to conceptualize the suffering of the people of the 'global South' due to the impact of climate change would be to invoke human rights. When they are embedded in laws or treaties, human rights are enforceable protections that generate obligations on the part of others. So declarations and conventions like the Universal Declaration of Human Rights and the European Convention on Human Rights can serve an important role when campaigners invoke them to make claims on their own behalf or to seek redress for parties whose rights have arguably been infringed. However, as McKinnon points out, rights can conflict, and the delivery of one right can clash with the recognition of another (McKinnon 2022: 41).

This problem of priorities can be resolved if we agree with Henry Shue that priority should be given to 'basic rights', which turn out to be highly relevant in matters of climate justice. Basic rights are rights in the absence of which no other rights can be exercised, and include the rights to life, means of subsistence, and freedom from violence and torture. No one is to be allowed to fall below the level of treatment that these rights enjoin (Shue 1996: 18). When basic rights are applied to the victims of climate injustice, the message becomes clear that the people of developing countries must be assigned priority because they 'face disproportionate threats to their basic rights' (McKinnon 2022: 42). And this, in turn, gives rise to imperatives to mitigate greenhouse gas emissions and also to support forms of adaptation to the presence of increased levels of greenhouse gases that are compatible with sustainable development and with the UN Sustainable Development Goals of 2015. Besides, where poverty prevents the fulfilment of basic rights and blocks climate action, climate justice involves nothing less than assistance from developed countries to introduce sustainable development, to eradicate poverty and thus to facilitate climate action.

While this approach supplies a viable route to climate advocacy of an ethical kind, it cannot be assumed that rights are morally basic. Many philosophers regard rights not as fundamental axioms but as conclusions (Dworkin 1977) and their justification as lying in the overall benefits of the rules that enshrine them being observed (Hare 1981: 147–68) or in the satisfaction of the basic needs that they uphold (Attfield 2019 [1995]: 142–4). These approaches remain consistent with appeals both to declarations, like the Universal Declaration of Human Rights, and to interpretations in which basic rights assume priority. Thus they supply grounds that are more secure for the same imperatives to mitigate emissions, to assist adaptation to emissions that are beyond retrieval, and to assist the introduction of sustainable development. The same grounds also support rights to compensation for those who have suffered 'loss and damage' from historical and recent emissions (McKinnon 2022: 99), a theme to which we return later in this chapter.

Human victims of climate injustice: climate refugees

Environmental refugees have been defined by Essam El Hinnawi as 'those people who have been forced to leave their traditional habitat, temporarily or permanently, because of a marked environmental disruption (natural or triggered by people) that jeopardized their existence and/ or seriously affected the quality of their life' (El Hinnawi 1985: 4). In 2005, the leading environmentalist Norman Myers revised an earlier estimate of his of the number of environmental refugees to a total of 200 million people (Myers 2005). More recently, Catriona McKinnon has estimated that by 2050 between 25 million and one billion people will be displaced by climate change, most of them in developing countries (McKinnon 2022: 49). Currently, because of the narrowness of the definition of 'refugee' in the 1951 Convention Relating to the Status of Refugees, environmental refugees have no standing as such in international law (except for those who happen also to be internally displaced persons) (Westra 2009: 4–5).

Most, if not all, environmental refugees are also climate refugees whose lives have been disrupted by extreme weather events or by sea-level rise. Catriona McKinnon cites the example of the islanders of Kiribati in the Pacific Ocean. The atolls that comprise Kiribati are on average only 2 metres above sea level, and are gradually disappearing beneath the waves. By 2100, no Kiribati atolls are likely to exist above water. The Kiribati government has bought an island in Fiji, to which a number of their people have moved. Other ideas for survival include building sea walls, constructing floating islands on which people could live, and dredging materials from the bottom of a lagoon to raise the height of the islands and thus avoid submersion (*Guardian* 2020; McKinnon 2022: 46).

The people of Kiribati are among the world's indigenous peoples, who comprise some 5 per cent of the global human population (McKinnon 2022: 43), or towards 400 million people. This means that, like other indigenous people, whether of Oceania or of other continents, their culture is closely bound up with their ancestral territory, for example, with dances in which they mimic the local birdlife (ibid.: 46). And this in turn means that, even if they were offered a new homeland, there are few substitute territories where their culture could be preserved intact. Alternative proposals for possible forms of compensation will be considered shortly. For there is clearly a case for compensating peoples who are being deprived of their existing homelands almost entirely through the impacts of the actions of other people and of other countries. But, importantly, the case for compensation also applies to other displaced people, whether indigenous or not.

There is a particularly strong case for compensation for the peoples of states whose entire territory is disappearing through climate change, which Cara Nine calls 'ecological refugee states' (Nine 2010). Nine's argument rests on the proviso of John Locke (1632–1704) for property rights being legitimate, namely that this depends on there being 'as much and as good' left for others to acquire similar rights. She goes on to remark that in the modern world there is no inhabitable land that is not the territory of one country or another. Hence, if Locke's reasonable proviso is accepted, states with viable territories have obligations 'to allow reasonable access to their territory to the ecological refugee states' (Nine 2010: 366). Even if we do not accept Locke's proviso, it could reasonably be argued on the basis of the right to life of the inhabitants of these states, and of these states' right to survival, that other states with viable territories have obligations of the kind that Nine advances.

As Nine concludes, this suggests that other states might have to relinquish some of their territory to meet this obligation. But this proposal has problems, since the majority of territories are occupied already. There are islands that have been abandoned (for example, in the Hebrides and the Scilly Islands), but they were mostly abandoned because of the difficulty of sustaining life in their climates and at a distance from hospitals, schools and supplies that were not locally available (Attfield and Clutterbuck 2014).

Another possibility is that states should allow the people of ecological refugee states the simultaneous use of some of their territory, where the refugee people (often, like the people of Kiribati, people who live by fishing) have different uses for territory from those of the current occupants. But this may well be an unstable possibility since both the resident population and the newcomers would need to use the same roads, schools and hospitals, and sovereignty would lie with whichever community was in charge of these facilities and of the taxes needed to pay for them (Attfield and Clutterbuck 2014).

A further proposal is that dispossessed people, whether from ecological refugee states or not, should be awarded passports by the international community (perhaps through

United Nations agencies) that would grant them freedom of movement, that required states where they arrived to grant citizenship to the holders of these passports (Heyward and Ödalen 2016). This proposal could serve as a remedy for the injustice of displacement or of territorial dispossession, and it was put forward as a 'second-best' solution in a situation where justice requires redress, but full redress (involving the granting of new sovereign territory) is impossible. As McKinnon comments, there is a danger with this non-territorial solution, as there also is with territorial ones, that the states where climate refugees might wish to relocate could be ones already suffering from climate damage (McKinnon 2022: 48–9). For this eventuality, the proposal, it might be suggested, should additionally provide for financial compensation for this new burden to be paid by developed countries to the states where the refugees decide to settle. If by 2050 the number of displaced people approaches anything like McKinnon's estimate, then this additional proposal could involve considerable expense; but justly so.

Where possible, it is desirable that submerging states should manage eventually to hold at least one island or territorial enclave where their flag can be raised and national passports issued as a symbol that their sovereignty has not been lost together with their territory (Attfield and Clutterbuck 2014). In the case of Kiribati, the island that has been purchased from Fiji might serve this purpose, even if most of the former population takes up residence in distant cities such as Auckland (New Zealand is generous in hosting dispossessed Pacific refugees), Sydney or Melbourne, or on the other islands of Fiji.

Yet for the much larger number of refugees, displaced by flooding or heatwaves or drought, whether in Africa, Asia or Latin America, justice requires an ampler solution. The World Bank has given the following estimates of the numbers of internally displaced people: 86 million in sub-Saharan Africa; 40 million in South Asia; and 17 million in Latin America (McKinnon 2022: 49; Rigaud et al. 2018). Internally displaced refugees in theory have their own government to turn to, but the governments of such countries frequently have inadequate resources and need international assistance, which could and should be funded by an international climate-change-related Loss and Damage Fund, such as the one mentioned in the final paragraph of this section.

However, millions more climate-change refugees are forced to cross borders for the sake of survival. Those states whose prosperity is largely due to greenhouse gas emissions should accept responsibility for these people's care and resettlement on the basis of both their own historic responsibilities and their current ability to pay; for climate justice requires nothing less. (Ability to pay is discussed later in this chapter.)

One proposal for the organization and funding required is the scheme of Frank Biermann and Ingrid Boas for a 'Protocol on the Recognition, Protection and Resettlement of Climate Refugees', to be hosted by the United Nations Framework Convention on Climate Change (UNFCCC), and funded by a 'Climate Protection and Resettlement Fund' (Biermann and Boas 2008; McKinnon 2022: 50). Nothing short of this proposal is likely to be commensurate with the problem, even if it can be supplemented by proposals to harness the collective agency of groups of climate refugees themselves.

In particular, Alex Arnall, Christopher Hilson and Catriona McKinnon have argued for the capacities of groups of affected people to be encouraged by local and national governments and by international bodies to develop solutions of their own, albeit living across frontiers and away from their homelands (Arnall, Hilson and McKinnon 2019). McKinnon adds that the problems are so great and widespread that it is 'highly likely that communities will need to play a central role in organizing themselves and advancing their own claims . . . rather than relying on central governments and authorities' (McKinnon 2022: 50). Self-help on the part of refugee communities is likely to be an indispensable part of the process of resettlement, and efforts to encourage this would also express recognition of the human dignity of the affected people. This said, the resources required are unlikely to be available in the absence of a fund such as that proposed by Biermann and Boas, and of their proposed internationally coordinated organization, placing pressure on the states where climate refugees arrive to participate, for example, by making available land, infrastructure and/or relevant education and training.

The current treatment of climate refugees falls far short of all this. Granted the requirements of justice, this treatment is to be deplored. Until or unless a Climate Protection and Resettlement Fund is set up, it would be appropriate for better provision for climate refugees to be funded in the first instance from the newly established international climate-related Loss and Damage Fund, agreed at COP27, the Conference of the Parties held at Sharm el-Sheikh in November 2022.

Non-human victims of climate injustice

But moral standing is not restricted to human beings. This is well argued by Cripps in her chapter 'Beyond Humans' (Cripps 2022: 73–93). Most ethicists now accept that at least sentient animals matter, morally speaking, and should not be made to suffer unnecessarily; this is accepted not only by the founder of the Animal Liberation movement, Peter Singer (Singer 1976) but also by such mainstream ethicists as Geoffrey Warnock and Richard Hare (Hare 1996; Warnock 1971).

Cripps argues (like Singer before her) that if all human beings have moral standing, then many non-humans with comparable capacities must have this standing as well. She also points out that those who deny moral standing to all non-humans, if consistent, cannot avoid denying it to many human beings too (Cripps 2022: 76–8). The question

'Can they suffer?', raised by Jeremy Bentham, suggests, as Bentham held, that the law (as well as morality) must afford protection to what he called 'sensitive beings' (ibid.: 78). This stance is now widely known as 'sentientism', and maintains that 'like interests should be given like consideration', whether they belong to a human being or to a non-human animal.

Yet creatures with interests are not limited to sentient beings. Many other creatures have capacities for growth natural to their kind, for reproduction and for self-maintenance. Some are animals lacking the kind of nervous system that may be needed for sentience, such as insects. Others are plants, with their characteristic capacity for photosynthesis, and others again are funguses, with their capacity to consume dead or dying organisms. All of them have a good of their own and are capable of being benefited. And this, as Kenneth Goodpaster has argued, gives them moral standing; benefiting is so central to morality that moral standing arguably belongs to any organism capable of being benefited. So moral standing attaches not only to sentient creatures but also to non-sentient animals and to funguses and plants (Attfield 2005; Goodpaster 1978). As Donald Scherer argues, we would seek to preserve any planet bearing life, as opposed to ones without it; and his disclosure reveals our acceptance of the moral standing of all living creatures (Scherer 1983). This is the ethical stance called 'biocentrism'.

As Goodpaster argues, a biocentric stance does not imply that all creatures must be treated alike, let alone equally. Different creatures can carry different degrees of moral significance, depending on their different ranges of capacities (Goodpaster 1978). Granted that equal interests should receive equal consideration (Singer's axiom), it does not follow that farm animals should be granted powers of which they are incapable, such as the right to vote, or that in a drought the last drops of water should be equally shared between plants, livestock and human beings. The ampler powers of most human beings mean that humans should often receive greater consideration and respect. Yet where capacities are shared, such as the ability to suffer, their bearers should be spared avoidable ill-treatment likely to cause suffering, whether they are human or not.

It is sometimes suggested that species and ecosystems matter in themselves and should also be recognized as having moral standing (Cripps 2022: 83–5). This is the stance known as 'ecocentrism'. But species (as opposed to their members) do not have a good of their own, and nor do ecosystems. This point has recently been explained in greater detail by the biocentrist John Nolt (Nolt 2015). The importance of species and of ecosystems lies in the value of the flourishing of their members, both present and future; and this is why they should be preserved.

How then can we account for the widely recognized distinctive importance of not exterminating the last few members of a species? We can account for this because this action eliminates the value that would or could have been present in the lives of all the future members of the species, which are (through the same actions) now prevented from coming into being. Yet, for these same reasons, both species and ecosystems are of high but indirect ethical importance, and should be preserved partly for the sake of their individual members (present and future) and partly for the sake of the other creatures, human beings and sentient animals included, that depend on their continuing survival and on the 'ecosystem services' that they supply. And the same can aptly be said about the biosphere as a whole, on which we and all other species depend.

Relatedly, the extreme weather events generated by climate change strike at billions (probably trillions) of non-human creatures (see chapter 3). Droughts, for example, are widely causing the starvation of cattle and of goats in places like the Horn of Africa, while wildfires burn to death many wild creatures, including both animals and plants. The strong winds caused by storms and hurricanes destroy whole hectares of forests, together with the animals that depend on them. There again, the kinds of floods associated with climate change (like the floods of October 2022 in Pakistan) cause livestock and wild animals alike to drown, and often destroy entire ecosystems, such as those of rivers and wetlands, together with all the living creatures, animal and vegetable, that comprise them.

Besides, the acidification of the oceans resulting from climate change is undermining the ecosystems of many coral reefs (as chapter 3 attests), and this in turn involves the death of both the corals and of the countless fish, crustaceans and other wildlife that depend on them. And the erosion of coastlines due to rising sea levels causes further harm to the wildlife of coasts. There will, at the same time, be some creatures that benefit, but it is implausible that the value that accrues from this outweighs the losses when cliffs, dunes and former inter-tidal wild communities disappear under the waves.

Granted the moral standing of all these plants and animals, it follows that it would be wrong for human beings with the ability to do so *not* to attempt to preserve them. In other words, we have obligations in their regard. It is widely maintained that obligations of this kind are not obligations of justice, not least by John Rawls (Rawls 1972). But this verdict is not surprising in a philosopher who makes justice depend on what rational parties would agree to in a fair bargaining situation, since such a situation excludes non-humans (and implicitly their interests) from the outset. More to the point, it is also widely recognized that when the vital needs of animals, such as freedom from a painful death, clash with the trivial interests of human beings (including those to whom we owe obligations of justice), the interests of the animals should be prioritized, and that this is only fair. This shows that we implicitly take it for granted that non-humans are included in the sphere of justice. The case for the treatment of non-human creatures being a matter of justice has been argued in greater detail elsewhere (Attfield and Humphreys 2016, 2017). This case makes it clear that non-human creatures are to be included among the potential victims of climate injustice, and that many of them are actual victims.

But does this make any difference in practice? Bryan Norton has argued that in environmental matters, anthropocentric and non-anthropocentric stances generate exactly the same policies (Norton 1991), and this might well make any focus on non-human needs and interests redundant. The first reply is that, once we recognize that non-human interests matter, we have a much stronger case to preserve non-human habitats, even where human interests support this as well. What is at stake is not just human life but the future of life on our planet.

If, however, the challenge is to identify cases where non-human interests require policies that human interests alone do not, or to which human interests would assign no more than a low priority, we need to reflect on the human treatment of zones and regions which are uninhabited, or else sparsely inhabited, by human beings. These include thick forests, deserts and the deep oceans, and also the upper regions of the atmosphere. Many of these zones and regions harbour wildlife, and some of them ecosystems. For such zones and regions, a non-anthropocentric stance, such as biocentrism, makes efforts to preserve these living creatures and their ecosystems imperative, while an anthropocentric stance does this only for cases where human interests happen to be at stake, which is far from all of them.

Here is an example. It has come to light that there are fish living in a lake of fresh water two miles beneath the surface of the Antarctic ice shield. While human interests (other than scientific curiosity) are indifferent with regard to their survival, human actions are capable of undermining their existence, either through mining for rare materials or through causing the ice shield to become thinner and their isolated ecosystem to become linked to that of the Southern Ocean, and thus salinized and destroyed. Biocentrism here urges policies of leaving their habitat alone, whereas anthropocentrism implies nothing of the sort. Even sentientism is inconclusive in this matter, because the sentience of fish is a controversial issue. Yet if we care about the survival of all forms of life on our planet, we should uphold the policies prompted by biocentrism. Fortunately, the Antarctic Treaty of 1959 prevents the militarization of Antarctica, while the associated Environmental Protocol precludes actions liable to pollute it or harm its wildlife (Secretariat of the Antarctic Treaty 1991).

One further example illustrates the practical differences between the implications of different ethical stances. Scientists believe that some of the moons of Jupiter and of Saturn may be capable of supporting life, albeit life of some primitive kind. A biocentric stance implies in this context that great care should be taken in space exploration, particularly where the extraction of minerals for terrestrial use is envisaged. However, anthropocentrism has no such implications beyond the safeguarding of those moons for the sake of scientific research, and this could well prove compatible with some amount of mineral extraction. Yet most people who reflect on issues such as this one would favour much greater care being taken with these moons than with uninhabitable moons and planets. The Swedish philosopher Erik Persson has attempted to uphold the view that this is the best policy on the basis of sentientism (Persson 2008), and has subsequently advised NASA accordingly; but since these extraterrestrial life forms are unlikely to be sentient, it is difficult to see how this can consistently be achieved, except perhaps through introducing an unusually strong form of the Precautionary Principle. The view that moons harbouring life (of a probably non-sentient kind) require great care prior to mining being inaugurated there clearly presupposes a biocentric stance, which those who hold this view turn out to have held all along.

However, the preservation of extraterrestrial life is only relevant to terrestrial climate policies if our implicitly biocentric intuitions about space exploration are allowed to mould our overall ethical stance, and the resulting stance is then applied to the overheating planet on which we live. It is on Earth that non-human creatures are known to be the victims of climate change, and, if the above reasoning is valid, can be known to be among the victims of climate injustice. Thus when we consider the shape that a world of climate justice would take, we need to consider their interests seriously, alongside the other victims discussed in previous sections, preventing their exploitation and protecting their habitats and their biodiversity (the rapid decline of which has already been depicted in chapter 3).

William MacAskill has argued, on the contrary, that the average well-being of wild non-human creatures is probably negative, or at least not positive enough to count in favour of fostering their continuing existence. He accepts that their well-being matters, but in view of his findings about the low probability that their lives are (on average) more than marginally worth living, welcomes the continuation of human expansion into their habitats (MacAskill 2022: 211–13 and 312, n. 78). However, he appears to be taking into account sentient animals only, thus omitting the well-being of the vast majority of non-human creatures, and their immense total value (as understood from a biocentric perspective). There is also reason to resist his view that lives (human or non-human) that are only marginally worth living are not valuable enough to be worth being continued, or for their species to be preserved into the future. I have argued elsewhere that all lives with a positive balance of well-being supply reasons for their continuation or preservation, in reply to Derek Parfit's view that there is a critical level beneath which well-being does not count as a reason for any kind of action at all (Parfit 1984), a view that MacAskill seems to endorse. Readers interested in that debate are referred to my earlier books (Attfield 2019 [1995]: 156–7; 2020 [1987]: 159–62).

In any case, MacAskill has enough confidence that the combined future prospects for human beings and non-human creatures are sufficiently high for policies of resisting global heating to be strongly desirable (2022: 220, 227). That means that his overall verdict on morality probably aligns him after all with many of the verdicts on climate justice expressed in this book (as the passage from MacAskill cited in the coming section appears to confirm).

Responsibility for climate justice

Climate justice is attracting increasing attention and has become the focus of an international movement (Jafry 2019). It involves, among other concerns, the mitigation of greenhouse gas emissions, the growth of which needs not only to be halted but actually to be reversed. The goal is to reduce emissions so that the climate stabilizes at no more than 1.5°C above pre-industrial levels. In this connection, the phrase 'net zero' is used of systems that are carbon neutral, but that phrase is prone to generate controversy that need not be further considered here.

The reasons for the 1.5°C target are that if emissions exceed this level, coastlines will be severely flooded, small islands like Kiribati will be submerged and extreme climate events will become even more extreme. There are also serious dangers that tipping points will be reached and that a sequence of climate systems will collapse (see chapter 2). Avoiding these disasters is vital for the interests of humanity (present and future) and for those of the rest of life on our planet as well. The ethical case could hardly be stronger.

Climate justice also involves the adaptation of all countries to the raised levels of greenhouse gases that have already been emitted (and irrevocably, at that). This adaptation will require huge adjustments to the infrastructure of most countries, with protections against both flooding and droughts, storms, heatwaves and wildfires. So vast is the scale of the needed changes that most developing countries are in no position to fund them in full; the funding needs to come from elsewhere.

Climate justice also calls for compensation to those who have suffered and are suffering from climate change from those who have caused it and can afford to pay to alleviate this suffering. As we have seen, the victim countries are often those whose emissions have been minimal, and they should be recompensed by those whose prosperity is due, in some part, to their past and present emissions. The amount needed to cover this 'loss and damage', according to BBC reports about COP27 in November 2022, amounts to US$3 trillion annually, most of which will have to be funded by the more developed countries.

Who, then, should fund this trio of projects: mitigation, adaptation and compensation? (For this triple focus on kinds of climate action, see Cripps 2022: 98.) Three bases have been suggested: that 'Polluters Pay', that the beneficiaries of climate change should pay, and that the responsibility rests with those able to pay (Cripps 2022: 109–21). These are overlapping categories, but their divergences warrant a brief discussion.

The principle that polluters should pay bears the hallmarks of 'natural justice'; those who have caused the problem should pay to remedy it. However, some polluters may no longer have the resources to pay and, more importantly, many polluters have either died or ceased to exist. Who is to be held responsible for the pollution from the Austrian Empire or the Ottoman Empire, or from the former Yugoslavia or the former Czechoslovakia? This principle has the merit of including historical pollution, but the defect that not all historical polluters are available to pay.

As for the idea of beneficiaries being expected to pay, this will include those northern countries that have become more fertile with climate change, as well as those whose wealth developed through industrialization. Yet there could be beneficiaries that are unable to pay, whether through natural disasters or, as in the case of Ukraine, through warfare.

This leaves those with the ability to pay, whether they were historically responsible for greenhouse gas emissions or not. This could seem unfair until it is recognized that all countries are currently contributing to such emissions. The principle that those able to pay should bear the burden of funding climate action also includes wealthy corporations and individuals, whose contributions would be made either through voluntary contributions or through taxation. This seems much the best principle, as long as those who have contributed most to pollution are recognized as obliged to contribute most, provided that they are able to do so.

To bring some concreteness to these somewhat abstract points, what those able to pay should be expected to pay for includes the reduction of their own greenhouse gas emissions (for example, through moving to renewable energy generation) and the adaptation of their own infrastructure (to avert the worst effects of extreme climate events). They should also contribute large sums to assist the mitigation and adaptation processes of poorer countries, and make payments into the newly agreed fund to remedy their loss and damage. All countries will need to contribute to their own mitigation and adaptation, and some of the costs of rectifying loss and damage, but developing countries cannot be expected to carry these burdens unaided. The contributions of developed countries turn out to have lagged behind even their own previous undertakings and need to be considerably increased if the impacts of coastal flooding and of widespread extreme climate events are to be remedied.

Before we move on, it is not only gratifying but positively enlightening to cite the perspective of MacAskill in the matter of the introduction of clean technology and of doing so in an innovatory manner. MacAskill writes as follows:

some actions make the longterm future go better across a wide range of possible scenarios. For example, promoting innovation in clean technology helps keep fossil fuels in the ground, giving us a better chance of recovery after civilizational collapse; it lessens the impact of climate change; it furthers technological progress, reducing the risk of stagnation; and it has major near-term benefits too, reducing the enormous death-toll from fossil-fuel based air pollution. (MacAskill 2022: 227)

MacAskill's final remark coheres well with the passages on the air pollution crisis in chapter 3 above. This extract also alludes to the fear that civilization as we know it could in theory collapse, and that in that event our struggling distant descendants could benefit if some fossil fuels were still in the ground, in case technology has to be reconstructed again from scratch. Here it can be remarked that large-scale resort in the immediate and near future to clean technology is one of the likeliest ways within the power of this and coming generations to prevent civilizational collapse from taking place.

The shape of climate justice

A different basis for both entitlements to emit greenhouse gases and for funding their mitigation is that of Contraction and Convergence (Meyer 2005), a proposal that I used to support (Attfield 2014 [2003]: 207–12; Attfield 2015 [1999]: 208–11). In this system, the total of allowable emissions would be gradually distributed to countries in proportion to their population, and countries seeking to emit more than their quota would need to purchase emissions quotas from those not using their own in full. The strengths of this plan included its emphasis on human equality and its implicit rejection of the assumption that those who had grown rich through polluting in the past could be regarded as free to continue in the same way ('grandfathering').

However, as emissions grew, the availability of under-used quotas was likely to dwindle. There again, no provision was included in this system for funding to protect wildlife and habitats from destruction. There was also a more serious and more fundamental objection, which is that this scheme disregarded historical emissions and focused on current ones instead. This objection was open to the reply that it was not clear that emissions of (what we now call) greenhouse gases were having a greenhouse effect until around 1990, and that historic emissions from before that stage could not fairly be blamed on those who had caused them. But as greenhouse emissions continued across the subsequent decades, it became unreasonable to disregard emissions of the post-1990 period. Besides, the ignorance of polluters prior to 1990 is inadequate completely to block the case for compensation. All this meant either that the plan for equal entitlements needed to be modified to reduce the entitlements of countries responsible for post-1990 emissions, and to make room for compensation, or that some altogether different basis was required.

Such a different basis appeared to be available in the proposal for Greenhouse Development Rights. This proposal makes provision for the development of developing countries, in addition to mitigation and adaptation on a global scale. In its original version, Paul Baer, Tom Athanasiou, Sivan Kartha and Eric Kemp-Benedict proposed a system of global taxation to which everyone with an income greater than the average for Spain would contribute (Baer et al. 2008). One strong merit of this proposal was that wealthy people in developing countries would be expected to contribute, instead of being exempt. The ethical case for this proposal is so strong that several charities such as Christian Aid have given it their support.

But there is also an ethical and political case against it. The proposal involves establishing an international authority empowered to raise taxes in every country, and to do so fairly and impartially, and then to administer its distribution to all countries in need of mitigation, adaptation or development. Such a system would depend on this international authority, and all its local agencies, being trustworthy and free of corruption, whether the local agencies were national governments or representatives of the international organization. But the prospects of this being agreed and then implemented fairly and impartially are so slight that the ethical case for agreeing to it is weak; and the political objections, rather like the objections to a world government, are overwhelming. Besides, this proposal takes inadequate account of historical emissions. Yet the central problem with this proposal is that its objectives are too broad, and that its operation would be unwieldly and fraught with debilitating problems.

Hence it is ethically preferable for mitigation and adaptation to be organized at national level, and for incoming contributions, to be supplied by developed countries, to allow developing countries to achieve this (together with the alleviation of poverty which, as we have seen earlier in this chapter, is a prerequisite) to be supplied by developed countries. The developed countries would at the same time contribute to a Loss and Damage Fund to prevent and/or remedy the impacts of coastal flooding and of severe climate events, including the severe climate events of recent decades. At the same time, the nationally determined contributions (NDCs) made at previous Conferences of the Parties need to be both observed in full and enhanced sufficiently to make the target of limiting the rise of average temperatures (above pre-industrial levels) to 1.5°C attainable. Currently, these NDCs are more frequently disregarded than observed; the focus of campaigners convinced that attaining this target is ethically imperative should be on these NDCs being both observed and ratcheted up so that this target can be attained.

The various funds released through enhanced NDCs and contributions to the Loss and Damage Fund would also have to be used to compensate the victims of climate change. Funding for dispossessed refugees and dispossessed countries should become available from these sources, whether this was officially called 'compensation' or otherwise described; this should include funding for the education and training of these people to enable them to resettle and rebuild their own futures.

There again, funding for developing countries should be drawn from these funds to facilitate their own mitigation and adaptation and, as a preliminary to this, to equip them to invest in these measures, rather than spend their income on debt repayments. Towards fifty developing countries are reported to be on the brink of insolvency (Harvey 2022b). But countries in such a vulnerable situation are in no position to invest in renewable energy or in the infrastructure needed to adapt to new levels of greenhouse gases, or to protect their own forests and biodiversity. International assistance is urgently needed, not in the form of loans, which would deepen their debt, but as grants.

Further, the protection of forests and of oceans should also be underwritten from these same funds. The rate of deforestation may have slackened in recent years, but it needs to be halted and reversed by reforestation if tipping points are to be avoided and if greenhouse emissions are to be sufficiently mitigated. Techniques are becoming available to renew coral reefs, but this activity too requires international funding. The case for all this consists in part in the good of non-human creatures, but human needs amply justify it as well.

In addition to the contributions of governments, corporations, non-governmental organizations and local authorities have a large part to play in greening their economies and societies, whether through reducing emissions, introducing sustainable forms of production, agriculture and energy generation, and/or through rewilding local areas. In these ways, contributions can be made both to the global climate and to reducing local levels of air pollution, thus improving human health. There is also large scope for individuals and households to adopt more climate-friendly lifestyles, reducing their emissions, their travel and their consumption of meat, offsetting their travel (see chapter 4), planting trees and campaigning for climate action to be taken by their governments.

While individual citizens lack the power to enact global change, their actions can bear witness to the changes that are needed, and their campaigns can make a significant difference. Although it is late in the day, it is not yet too late to strive to 'keep 1.5 alive'. Individuals and local groups who do so can serve as a vanguard, living locally the kind of life that is urgently needed globally.

Recommended reading

Attfield, Robin. 2014 [2003]. *Environmental Ethics: An Overview for the Twenty-First Century*, 2nd edn. Cambridge: Polity.

Attfield, Robin. 2015 [1999]. *The Ethics of the Global Environment*. Edinburgh: Edinburgh University Press.

Biermann, Frank and Boas, Ingrid. 2008. 'Protecting Climate Refugees: The Case for a Global Protocol'. *Environment: Science and Sustainable Development* 50(6): 8–17.

Brown, Donald A. et al. 2005. *White Paper on the Ethical Dimensions of Climate Change*. Rock Ethics Institute, Pennsylvania State University.

Cripps, Elizabeth. 2022. *What Climate Justice Means and Why We Should Care*. London: Bloomsbury Continuum.

Goodpaster, Kenneth E. 1978. 'On Being Morally Considerable'. *Journal of Philosophy* 75: 308–25.

McKinnon, Catriona. 2022. *Climate Change and Political Theory*. Cambridge and Hoboken, NJ: Polity.

Nolt, John. 2015. *Environmental Ethics for the Long Term: An Introduction*. Abingdon and New York: Routledge.

Singer, Peter. 1976. *Animal Liberation: A New Ethic for Our Treatment of Animals*. London: Jonathan Cape.

Westra, Laura. 2009. *Environmental Justice and the Rights of Ecological Refugees*. London and Sterling, VA: Earthscan.

Further reading

Attfield, Robin. 2020 [1987]. *A Theory of Value and Obligation*. Abingdon and New York: Routledge.

Attfield, Robin and Clutterbuck, John. 2014. 'Climate

Refugees, Disappearing States and Territorial Compensation'. *Proceedings of Philosophy, Yesterday, Today and Tomorrow Conference*. Singapore: Philosophy Yesterday, Today and Tomorrow Conference.

Baer, Paul, Athanasiou, Tom, Kartha, Sivan and Kemp-Benedict, Eric. 2008. *The Greenhouse Development Rights Framework: The Right to Development in a Climate-Constrained World*, 2nd edn. Berlin: Heinrich Böll Foundation, Christian Aid, Eco-Equity and the Stockholm Environment Institute.

Caney, Simon. 2012. 'Just Emissions'. *Philosophy & Public Affairs* 40(4): 255–300.

Jafry, T. (ed.). 2019. *Routledge Handbook of Climate Justice*. Abingdon and New York: Routledge.

MacAskill, William 2022. *What We Owe the Future: A Million-year View*. London: Oneworld.

Meyer, Aubrey 2005. *Contraction & Convergence: The Global Solution to Climate Change*. Totnes: Green Books.

Nine, Cara 2010, 'Ecological Refugees, States' Borders and the Lockean Proviso'. *Journal of Applied Philosophy* 27(4): 359–75.

Persson, Erik. 2008. *What is Wrong with Extinction?* Doctoral thesis, Lund: University of Lund.

Scherer, Donald. 1983. 'Anthropocentrism, Atomism and Environmental Ethics', in Donald Scherer and Thomas Attig (eds), *Ethics and the Environment*. Englewood Cliffs, NJ: Prentice-Hall, 73–81.

6

Some Political Implications

Climate justice in a non-ideal world: international responsibilities

The outcomes of the two major United Nations conferences of 2022 were in both cases disappointing and at the
same time encouraging. The COP27 Climate Conference,
held in November 2022 at Sharm el-Sheikh, failed to make
substantial progress on the phasing out of fossil fuels
(retaining the 2021 COP26 formula that they were merely
to be 'phased down'), but agreed to establish a Loss and
Damage Fund to assist developing countries that have
suffered from the impacts of climate change, despite the
reluctance of several developed countries to approve this
fund. Then, in December 2022, the COP15 Kunming–
Montreal Biodiversity Conference agreed to protect 30 per
cent of land and 30 per cent of oceans for the preservation
of species, to adopt targets for species preservation, and
to set up a related fund due to reach £30 billion by 2030.
The agreed text also mentioned the importance of indigenous peoples in protecting biodiversity, and of curtailing
subsidies for the kind of 'development' that serves to destroy
wildlife and to foster global heating. Disappointingly, the
targets were voluntary, but the very fact of their being
agreed was in itself a major achievement. The agreement
was almost vetoed by the Democratic Republic of Congo,

but was declared to have been carried by the Chinese chair, while subsequent negotiations prevented its collapse (Greenfield and Weston 2022a, 2022b).

Subsequently, in March 2023, the United Nations agreed a High Seas Treaty to preserve 30 per cent of the oceans (BBC 2023a), as was related in chapter 3 above in the context of biodiversity preservation The extent to which this treaty will be implemented remains to be seen, but this agreement shows that the work of the Kunming–Montreal Conference is being taken forward.

These developments bear out that progress towards climate justice is not impossible, but that the international context of these efforts is very far from ideal. Recent writers have addressed the problems of achieving climate justice in a non-ideal world. This chapter concerns the political arrangements and policies that are most appropriate in such a world in the light of the ethical findings about justice with regard to the issues of the climate and related crises, emerging from the two previous chapters.

Roser and Seidel usefully discuss 'what to do if others do not pull their weight' (despite having agreed to do so), writing in the context of countries reducing carbon emissions and contributing to international funding to facilitate mitigation and adaptation on the part of poorer countries (Roser and Seidel 2017: 169–78). They suggest that where another party defaults on an agreement and no third-party interests are significantly at stake, it is generally legitimate simply to discard the agreement. But international climate agreements are different because the interests of large numbers of future people are at stake. So one's own country should consider making good the shortfall, since this is an issue of intergenerational justice, which can outweigh intra-generational justice.

Here it could be commented that, in cases of climate agreements, there is an issue of intra-generational justice as well, because the well-being of numerous human contemporaries is also at stake. Besides, this is an issue of inter-species justice, too; for the very continued existence of many species depends on compliance with such agreements and, when some parties default, on others stepping up with additional contributions. I am not suggesting, and nor, probably, are Roser and Seidel, that any one country should make additional contributions indefinitely whenever others default, since this would be unfair to the tax-payers of that country and would be unsustainable. But, like them, I would advocate not giving up at the first sign of defaulting by others, and to use diplomacy to attempt to restore ongoing multilateral compliance.

More recently, McKinnon has also discussed 'options in the case of climate failure'. As she relates, Climate Action Tracker tells us that most countries are very far from delivering on the promises they made in their NDCs (that is, their nationally determined contributions, made at the Paris Climate Agreement of 2015) (McKinnon 2022: 121). As McKinnon explains, this has led to a range of kinds of 'non-ideal theory'. Non-ideal theory contrasts with 'ideal theory', according to which all parties agree progressively to mitigate their emissions and adapt their infrastructures to global heating, and also to contribute according to their means to the mitigation and adaptation of developing countries, and to appropriate compensation for loss and damage resulting from climate change, whether the term 'compensation' is used or not.

Non-ideal theory can involve devising incentives for 'pro-climate behaviour' (McKinnon 2022: 126), compatible with democratic systems. One example consists in carbon-trading schemes, whereby companies (for example, within the European Union) can purchase permissions to pollute. Such schemes, however, depend on suitable caps being set on total emissions, suitable prices being charged for licences to pollute, and suitable policies being put in place for the deployment of the resulting revenues. But historically, few, if any, of these requirements have been satisfied. Other forms of incentives to minimize emissions, such as carbon taxes, may fare better.

McKinnon also introduces the possibility of what she depicts as an 'ecocentric theory of value', and links with Richard Routley's Last Person thought-experiment (McKinnon 2022: 123–4). If the last person on Earth lays about him, destroying (say) a tree, even though neither he nor anyone else can benefit, most people hold that this behaviour is unjustified, thus disclosing that not all value is taken to depend on human well-being, and that some value is recognized to attach to the continued flourishing of non-human living creatures including trees (Routley 1973). This stance is in fact usually described as 'biocentrism', for which there is intrinsic value in the flourishing of all individual organisms, in contrast with the stance of ecocentrism, which ascribes intrinsic value to species and/or ecosystems, either as well or instead (Attfield 2014 [2003], 2015 [1999]; Nolt 2015). Indeed, biocentrism can supply a much more promising basis for non-ideal theory.

McKinnon distinguishes between reformist activism, which she regards as anthropocentric (focused exclusively on human interests) and radical activism (interpreted and represented as ecocentrist) (McKinnon 2022: 124). But this interpretation would confine radical activism to people influenced by ecocentric writers such as Arne Naess and Warwick Fox. In fact, many (perhaps most) environmental activists are either sentientists or biocentrists (see chapter 5 above), and thus reject anthropocentrism and its assumption that nothing but human interests matters. Once this is recognized, the prospects can be seen as much stronger for the 'ecological citizenship' that McKinnon implicitly commends (ibid.: 126–32), and for campaigns, both national and international, to uphold what was agreed at Sharm el-Sheikh and at Montreal in 2022, and to deliver what was promised at those and previous international climate conferences.

It should be added that, while far from ideal, the Kunming–Montreal Agreement (see p. 100), with its provision for preserving disappearing species, was implicitly far from anthropocentric. At least this provision goes far beyond what concern to preserve ecosystem services for humanity would warrant. Here we find traces of a non-anthropocentric (though not necessarily ecocentric) value theory surfacing in a United Nations conference in which nearly 200 nations participated. Despite the non-ideal world in which we find ourselves, this agreement embodies signs of genuine hope that the planet may yet be saved.

Some related policy issues

Policy issues and solutions pursuant to climate justice include those requiring international agreement, such as the policies accepted at COP27 and at the Kunming–Montreal Biodiversity Conference, and also national and local policies, which need not wait for international agreements. But some climate policies can be retrograde, depending on who is in power. Thus the policy of Jair Bolsonaro, then president of Brazil, to authorize massive deforestation of the Amazon in the name of development, was a threat to the global climate system, but it has happily been replaced by that of the new president Luiz Inácio Lula da Silva, inaugurated on 1 January 2023, who aims to preserve the Amazon forest (with 'zero deforestation') and to make compatible provision for the poor. During his first few days in office, he cancelled initiatives of his predecessor that had made land grabs easier, and gave new powers to the National Environment Council, and on his very first day he 'signed a measure creating the Amazon Fund, which works as a mechanism for foreign governments to help pay for preservation efforts' (Pagliarini 2023). Likewise, the new pro-climate government in Colombia promises to protect its portion of the same mega-forest far better than its predecessor.

There remains scope for follow-up action to ensure that what was agreed at Sharm el-Sheikh and at Montreal is implemented, not least on the part of the treasuries of the richer countries. All ministries of agriculture should be encouraging farmers to reduce methane emissions, for example by adopting the kinds of feed that lower methane outputs. Meanwhile, ministries of energy should be planning to phase out (and not just 'phase down') the mining and use of coal, and to leave fossil fuels of all kinds in the ground, while accelerating the introduction of renewable energy in all its many forms.

Where national policies are concerned, Saleemul Huq, director of the International Centre for Climate Change and Development, described (during a recent BBC Radio 4 programme) how Bangladesh has anticipated the need for adaptation by measures such as flood prevention and installing water-tanks on housetops to collect fresh water during monsoons. This has been achieved through domestic investment, prior to any international funding being agreed (Huq 2023). The Loss and Damage Fund could, if adequately supported, facilitate similar provision on the part of flood-prone countries like Pakistan.

For its part, the British government did well to award an honour to Alok Sharma, who chaired the (partly successful) COP26 Conference held at Glasgow in 2021; but less well to ignore Sharma's protests and approve recently a new coal mine in west Cumbria, United Kingdom. Despite the special case that has been made for the supposedly exceptional nature of this development, the British government is in this way undermining its capacity to urge coal-dependent countries like Poland and India to discard their dependency on coal and to move to renewable alternatives with minimal delay.

Similar verdicts are in place about the United Kingdom's issuing of licences for new oil and gas exploration in the North Sea and in the Rosebank 'oilfield' north-west of Shetland. For new oil and gas extraction is likely to lead to fossil-fuel production being unnecessarily extended across the decades when developed countries such as the United Kingdom should be leading the world into green technologies and into the complete abandonment of carbon-based energy systems. British connivance in the continuation of fossil-fuel energy generation contributes to the ongoing melting of the Greenland and West Antarctic ice sheets, which could make the attainment of a ceiling of 1.5°C all but impossible.

In the United States, the Biden administration has done well to secure the passage of the Inflation Reduction Act, which for the first time makes available billions of dollars towards a transition to a green economy. This measure was not flawless, as it also included subsidies for fuel companies. Indeed, the phasing out of such subsidies worldwide is needed if climate justice is to become a reality. However, this concession proved necessary to secure a majority in the Senate for the enactment of this legislation. Praise is also in place for counterpart legislation both in the European Union and in Japan.

What was said above about the proposed new coal mine in England applies also to the process of fracking for oil and gas, and to the extraction of oil from tar sands. These processes can appear tempting at a time of energy scarcity, but contribute blatantly to the crisis arising from greenhouse gas emissions and also to the despoliation of good and productive land, and in Canada to that of the ancestral territory of indigenous peoples. Indeed, the myth that gas-based energy is a form of green energy should be acknowledged for the lie that it is. Like gas and oil in general, these specific kinds of oil and gas must be left in the ground if climate justice is to be secured.

Nor is nuclear energy an acceptable green substitute. While it is not usually thought to contribute to greenhouse gas emissions, nuclear energy should be shunned while there is no known safe way of decommissioning discarded power stations or of storing nuclear wastes with half-lives of multiple thousands of years. The lingering radioactivity of these nuclear wastes is a clear example of the foreseeable impacts of current human agents on far distant human generations and of our responsibility to prevent them. The new dangers arising from warfare around the Zaporizhzhya nuclear power plant in Ukraine are an additional warning of the kind of near-catastrophe that arose in 1986 at the Chernobyl nuclear power station (also in Ukraine but before its independence) (World Nuclear Association 2022).

Besides, we may need to revise the standard view that nuclear fission does not contribute to global warming. A recent letter of early January 2023 in the *Whitehaven News* from Lakes Against Nuclear Dump quotes the Nuclear Decommissioning Authority as acknowledging that its total carbon footprint for 2019–20 was 1,046,950 tonnes of carbon dioxide equivalent (Lakes Against Nuclear Dump 2023). With emissions of this volume, the world would surely be a better place without power stations that pollute to this extent in the process of decommissioning, quite apart from their toxic legacy.

The likely exception to the unacceptability of nuclear power is power from nuclear fusion, which promises not to produce radioactive waste products, and which has recently emerged as perhaps genuinely feasible, as well as involving zero greenhouse gas emissions. However, the problems of scaling up reactions that generate an excess of energy outputs over inputs mean that this form of energy generation cannot be expected to be in use for several decades. Thereafter, it is possible that energy from nuclear fusion can supplement renewable energy and solve many of humanity's energy needs and problems. (This prospect has a bearing on proposals for climate engineering, discussed in the next section.)

A further salutary policy change would be withdrawals from the European Energy Charter that was agreed in 1991 and has 69 signatories (many from outside Europe) (European Energy Charter 2015). This Charter provides for companies to sue governments whose climate policies could curtail their activities or profits. But this manifestly inhibits national efforts at greenhouse gas mitigation, such as the phasing out of coal mines in the Netherlands and the banning of offshore oil drilling close to the coast by the government of Italy. Since the summer of 2022, a stream of countries has announced plans to withdraw from the related Treaty: Denmark, France, Germany, Luxembourg, the Netherlands, Poland, Slovenia and Spain. The pressure group Global Justice Now is campaigning for the United Kingdom to follow their example (Blaylock 2022). Indeed, it would be a major coup for climate justice if the remaining countries were to do the same. While the concerns of companies are often limited to the financial interests of their shareholders, those of democratic governments include the present and future well-being of their citizens, their other residents and the natural environments that they bequeath to coming generations. For these reasons, and in view of the ethical obligations of countries expounded in previous chapters, they would be well advised to follow the course indicated by climate justice, and allow this Treaty to become a dead letter.

Meanwhile, the human community needs a 'Climate Peace Clause', or a moratorium on the use of trade or investment rules in international agreements to challenge governments' climate policies. The time available for mitigation is short, and no time can be wasted on countering challenges to introducing and implementing the policies that are needed. The adoption of a Climate Peace Clause is advocated by the Sierra Club and the Trade Justice Education Fund. Such a moratorium could be declared unilaterally, for example at the World Trade Organization (Sierra Club 2021).

Climate engineering

Climate engineering was originally proposed as a supplement to greenhouse gas mitigation. But as time went by without any significant international agreement on mitigation, it began to be suggested as an alternative. Even now, with the Glasgow (2021) and Sharm el-Sheikh

(2022) Agreements in place (see the opening section of this chapter), some would maintain that their weaknesses, including the 'phasing out' of the use of coal being replaced by 'phasing down', indicate that climate engineering could soon appear to be needed.

There are two kinds of climate engineering: carbon dioxide reduction (CDR), and solar radiation management (SRM); and both kinds can assume either milder or stronger forms. Some of the milder forms of CDR include tree planting, already commended above in chapters 3 and 4, the cultivation of seagrass, and the preservation and restoration of peat bogs. Beds of seagrass are comparable in effectiveness to rainforests in sequestering carbon dioxide, while peat bogs, if left intact, continue to keep carbon sequestered. These forms of CDR are largely harmless, while the cultivation of seagrass in particular can restore some of the lost biodiversity of seas and oceans, just as the restoration of forests can preserve or restore some of the disappearing biodiversity of islands and continents. The main drawback of reliance on these forms of CDR is that they have a long 'lead-time' (Ott 2011), so the climate crisis could become overwhelming before they can make a sufficient difference.

The same verdict probably applies to attempts to keep carbon out of circulation by burying it in the form of biochar (charcoal made from biomass), a process which also enhances the soil. This process has a long history, not least in the Amazon region, which warrants new application in the context of the need to mitigate greenhouse gases. But its overall contribution would still be tiny in face of the vastness of the climate problem.

Another form of CDR is carbon capture and storage (CCS), in which the emissions of coal-, gas- and oil-fired power stations would be captured, before they can be discharged into the atmosphere, and stored underground. This project shows real promise of partially resolving the climate crisis, particularly if energy companies were required to adopt it. But the necessary technology for carbon capture has not yet been shown to be effective at scale. There is also the problem that if the carbon dioxide buried underground should leak out to any considerable extent, then it might prove to have been better not to even embark on this project in the first place. CCS should still be researched and pursued, but its introduction is unlikely to make a sufficient difference soon enough.

A similar comment is in place about milder forms of SRM, such as painting roofs white so as to reflect incoming solar radiation and thus prevent its absorption. This might be aesthetically displeasing, yet if widely and increasingly adopted it could limit some small proportion of incoming solar energy. But granted the scale and the urgency of the problem, such a scheme could not remotely make enough of a difference in time.

The limitations of these milder forms of climate engineering have led some to propose forms that would be stronger, bolder and more aggressive. A relevant stronger form of CDR would be to seed the oceans with iron filings, with a view to generating large blooms of blue-green algae, which could absorb many megatonnes of carbon. This project might be successful in reducing atmospheric carbon dioxide, but it could well contribute to undermining oceanic ecosystems and worsening biodiversity loss. It also seems inconsistent with the Kunming–Montreal Agreement of 2022 to make 30 per cent of the oceans reserves for wildlife, unmodified by human activity and technology.

Besides, it threatens to turn the oceans bright green (Ott 2011). This impact might appear worth risking but could disrupt the whole context in which many species, including humanity, have evolved. Could John Masefield have written 'I must go down to the seas again' if the sea itself were indistinguishable from a green soup (Masefield 1902)? Or would Edward Lear have written of 'The Owl and the Pussy-Cat' going to sea 'in a beautiful pea-green boat' (Lear 2023 [1871]) if the waves had by then become luridly green already? While these questions may seem gratuitous, they are intended to evoke the importance of preserving the context of the seascape in which our species and many others evolved. Resort to seeding the oceans with iron filings should, then, be deferred until there is no other solution to the climate crisis remaining. We are nowhere near that point at present.

Others have suggested a stronger form of SRM in which thousands of sulphate aerosols would be released into the stratosphere, with a view to incoming radiation being reflected back into space and not being allowed to enter our planet's atmosphere, as is described by Stephen Gardiner, a critic of such proposals (Gardiner 2011). But until sufficient effective forms of greenhouse gas mitigation are introduced, this process could well need to be continued for a long period. So it is important to consider possible side effects.

One possible side effect is chemical pollution. If the sulphate aerosols came into contact with the clouds of the atmosphere, precipitation could take the form of dilute sulphuric acid, to the detriment of all the communities and ecosystems on which it would fall. The acidification of the oceans (which is endangering coral reefs: see chapter 3) is already a side effect of increased carbon dioxide in the atmosphere; and the proposed release of sulphate aerosols could potentially make this problem considerably worse, as well as striking at all living creatures both on land and at sea.

There is also the problem of reversibility. Levels of greenhouse gases are continuing to rise, despite the various agreements about mitigation reached over recent decades (from Kyoto in 1997 onwards). If they were to continue to rise for some years after the aerosol-release process began, then a decision would need to be taken about whether to proceed with this process or to discontinue it. But discontinuation would be likely to produce a sudden rise in greenhouse gas levels in the atmosphere, threatening drastic impacts. Accordingly, the process of aerosol release could well have to be continued indefinitely. It might be preferable not to embark on a dangerous process which could effectively be irreversible, especially in view of the prospect that nuclear fusion may (within a few decades) obviate the need for climate engineering altogether.

Another problem is that the sky might well cease to be blue and become a milky-grey colour instead. But if a change of colour of the oceans could have far-reaching effects by changing the evolutionary setting of all oceanic and coastal creatures, a change of colour of the sky above our heads could generate yet greater impacts on all creatures everywhere, except those of the deep oceans. It is not just that poets like Thomas Traherne could no longer write of 'The skies in their magnificence' (Traherne 2023); the evolutionary context of (nearly) all life on our planet would have changed, perhaps irretrievably. Such experiments with our planetary home should not be undertaken unless there were no other resorts. But that is far from being the case.

Recognition of the crime of ecocide

One of the factors that explain non-compliance with the Paris Agreement on Climate Change, which agreed to 1.5°C above pre-industrial levels of greenhouse gas emissions as a desirable ceiling, is the lack of any international law criminalizing acts of large-scale destruction of the environment. Even at national level, in those countries that have environmental laws, enforcement is often unpredictable and lax. All this makes the playing field for energy companies and large emitters such as airlines and shipping lines uneven, allowing 'unscrupulous' companies to take advantage of legal loopholes, and giving them an advantage 'over more scrupulous competitors' (Roupé and Ragnarsdøttir 2022).

The proposed international legislation that would criminalize ecocide defines 'ecocide' as unlawful or wanton acts committed in the knowledge that there is a substantial likelihood of severe and either widespread or long-term damage to the environment being caused by these acts. Here 'wanton' means 'with reckless disregard' (for outcomes of that kind) (Stop Ecocide International 2022).

This proposed law would protect both the oceans and carbon sinks such as forests, seagrass beds and peat bogs. It would encourage the flow of finance towards renewable energy, and level the international playing field of investment by deterring companies and countries from investing in fossil fuels. For example, the plentiful subsidies for fossil-fuel extraction (mentioned earlier in this chapter) could well be curtailed and replaced by investment to protect the environment. Renewable energy generation currently amounts to 20 per cent of total global energy production, but governmental subsidies for fossil fuel retard its increase; the proposed legislation would be likely to remove this blockage and liberate the potential of renewables just when such expansion is most needed.

For time is short. The IPCC report on Mitigation and Climate Change, published in April 2022, asserts that emissions need to have peaked by 2025 and to be reduced by 43 per cent by 2030 if humanity is to have a chance of keeping global temperatures on a 1.5°C path. If we can make ecocide a crime in the Rome Statute (which governs the International Criminal Court), we may be able to combat climate change in time. Currently, there are four crimes recognized in the Rome statute: genocide, crimes against humanity, war crimes, and the crime of aggression. Adding ecocide as a fifth would be 'a comparatively rapid and straightforward process' (Roupé and Ragnarsdøttir 2022: 4–6).

'Making ecocide an international crime' (they add) 'would make top decision makers personally accountable for decisions that cause, or risk causing, mass environmental damage or destruction. This will stop many potentially harmful activities before they happen' (ibid.).

As the editors of *Ecocide Law* explained in a recent Schumacher Foundation webinar of November 2022, there is effectively no alternative to international legislation against ecocide. Since this law relates to people with power, rather than those without it, some of those people may attempt to resist the enactment of this legislation. Currently, many governments succumb to pressure from businesses to continue allocating subsidies. But the ecocide law would alleviate these pressures. In 1996, the proposed legislation was almost adopted, but three or four countries vetoed the measure. Since then, however, their views have shifted, and so the prospects for its adoption are favourable.

The process of adoption could take three years. But 'as soon as [an] Ecocide Law is on the horizon, it will begin to curb emissions from wanton fuel extraction.' This is because 'even the risk of the activities being deemed illegal will weaken the business case for them.' Business cases perforce take coming decades into account, and 'sustainable options will thus become more profitable and gain further momentum' (Roupé and Ragnarsdøttir 2022: 3).

Some of the likely impacts of the introduction of an Ecocide Law include governments transferring subsidies away from fossil-fuel companies to renewable sources of energy, thus accelerating the development of the latter. They also include deterring the many oil-extraction projects that involve risk of ecocide (as defined in the proposed legislation), and encouraging energy companies 'to adopt better approaches'. Another practice likely to be included is the protection both of carbon sinks and of large areas of the oceans that are currently outside national jurisdictions. The proposed law would not stop all mining but would require it to adopt responsible forms and 'more rigorous safety measures' (Roupé and Ragnarsdøttir 2022: 3).

As Alasdair Skelton, one of the advisers to the report's editors, commented at the Schumacher Institute webinar, those who support the introduction of an Ecocide Law should lobby for it with national parliaments and with local members of parliament, for example by visiting the surgeries of MPs. Many countries are currently considering this measure.

The authors of the report have added that its introduction is likely to modify corporate values (Roupé and Ragnarsdøttir 2022: 3). Here, however, it is important to elicit the ethical basis of the campaign to introduce ecocide legislation.

This legislation would contribute to alleviating both the climate crisis and the crisis of biodiversity loss through its impact on the mitigation of greenhouse gas emissions and on the protection of carbon sinks such as forests and of the oceans. It could well contribute to resolving the air pollution crisis too. But what makes the alleviation of these crises important is the value of the well-being of both current human generations and the generations of the foreseeable future, and of that of non-human creatures of this and of future generations as well. Acceptance of the moral standing of non-human creatures and of the intrinsic value of their well-being (as advocated in earlier chapters) strengthens this basis; nothing less than the survival and continuation of flourishing life on our planet supplies the ethical basis for the campaign to end ecocide.

No wonder that this campaign already has the support of Greta Thunberg, Jane Goodall and Pope Francis, as Monica Schülde (one of the editors of the report) informed the webinar. Corresponding national laws are likely to supplement the impact of the proposed new international law. Potential supporters, whether their profile is high or low, are encouraged to give public support to this campaign.

The moral responsibilities of companies

Most books and articles on environmental ethics and politics stress the responsibilities of governments and then of individual households. But a huge difference can be made by companies or corporations, some nationalized but most in the private sector. So it is companies that will here be considered first.

By no means are all companies environmentally irresponsible. Insurance companies in particular have added to pressures for society to move away from reliance on fossil fuels and towards a greener economy, not least because they are professionally involved with reducing risks of future harm and future damage. Green energy companies are playing an even more constructive role, as are the manufacturers of solar panels, electric storage batteries and electric cars; their orientation towards making profits makes them more rather than less inclined to take seriously the interests of their clients as well as of their shareholders. Companies can also contribute to the practices needed for adaptation, which Cripps lists as: 'insurance and education, adjusting infrastructure, finding ways to produce food using salt water, and developing early warning technologies for extreme weather'. Among further examples of adaptation, she includes 'a floating farm in Rotterdam' (Cripps 2022: 99). Many other examples could readily be given.

Yet the record of the petrochemical industry has seldom been a constructive one. It has recently come to light that Exxon scientists accurately predicted in 1977 the extent of the global heating of the present, and that nevertheless Exxon continued for decades to lobby against measures to mitigate greenhouse gas emissions. A spokesperson for Exxon, Todd Spitler, has declared that the company 'is committed to being part of the solution to climate change and the risks that it poses', and it is now involved in developing technologies like CCS and the use of hydrogen as a fuel. Yet for many years it questioned whether anthropogenic climate change was taking place at all, and whether there was any need to cut emissions, despite having access to the findings of its own researchers from the 1970s and 1980s onwards (Clark, Storrow and Harvey 2023).

Relatedly, Naomi Oreskes and Erik M. Conway have chronicled how such climate scepticism has been intertwined with the scepticism of those who denied the harmfulness of tobacco on behalf of certain tobacco companies. Such denialism continued, despite scientific consensus about global heating, into the early twentieth century, with other petrochemical companies participating, and extended to Europe and Australia as well as the United States (Oreskes and Conway 2010; Oreskes et al. 2018). Petrochemical companies need radically to change their operations to producing renewable energy and facilitating green transport, and to phase out their extraction of gas and oil as soon as possible. They should also discontinue their intensive lobbying at international conferences, such as the COP27 Conference at Sharm el-Sheikh.

Other corporations need to change more than their messaging. Cripps relates that in 2020 the (British) 'Advertising Standards Authority banned, as misleading, an advertisement describing Ryanair as "Europe's lowest fares, lowest emissions airline"', adding that in 'the same year, the Dutch Advertising Code Committee ordered Dutch airline KLM to change a campaign that over-emphasized the firm's use of biofuels' (Cripps 2022: 140). Firms intent on appearing green need to replace green-washing with a matching green performance.

Cripps further cites Oliver Lazarus as stating that '[t]he ten biggest US meat and dairy companies have . . . "lacked transparency about their emissions, lacked sufficient mitigation targets, or worked to influence public opinion on climate policy"' (Cripps 2022: 140; Lazarus, McDermid and Jacquet 2021). The immensity of the emissions of the factory farming industry was recently disclosed in a message from We Move Europe. It seems that 'a quarter of global emissions come from mass-scale factory farming'; manifestly, this 'model' of farming damages 'the planet and our health' (We Move Europe 2023). This is clearly a model that should be discontinued, quite apart from its exploitative treatment of the animals farmed.

One further example brings to light the difference that companies can make to the prospects for carbon mitigation. Livia Lie writes that Argentina is home to a deadly 'carbon bomb' 'because of the gas reserves in Vaca Muerta, an area rich in shale oil and gas deposit located in Patagonia. Fossil-fuel companies have been fracking in the region and polluting our planet.' Citing Peri Dias, she adds that 'If they exploit all the shale gas reserves in the region, it could take over 11 per cent of the carbon budget our planet has left to stay below 1.5°C of warming' (Lie 2023; see also Dias 2022). Unfortunately, the Argentine government is subsidizing these companies. So it is time to turn to the responsibilities of governments.

Responsibilities of governments

It is first worth saying that governments should not make matters worse with respect to greenhouse gas emissions. For the sake of the preservation of biodiversity and the prevention of excessive global heating, the Argentine government should discontinue its subsidies for fracking in Patagonia (see the previous section). But Argentina is far from the worst offender where fracking is concerned; both the United States and Canada should likewise desist. As Chris Skidmore MP has advocated, the British government should make greater efforts to decarbonize the British economy (BBC 2023c). Moreover, all the developed countries should abandon their subsidies to energy companies.

Yet not all subsidies are bad, and inaction by governments is sometimes a way of making matters worse. The emissions of the steel factory at Port Talbot in Wales amount to 2 per cent of UK emissions, and 15 per cent of the emissions of Wales, because of the large quantities of coke used to manufacture steel. This method of steel manufacture needs to be replaced by the introduction either of hydrogen as the fuel (already in use in Sweden) or of an electric arc process, fuelled by electricity generated renewably. This replacement would require government funding of several billion pounds, since the company would be unlikely to afford such a high investment (as discussed by Justin Rowlatt: BBC 2023d). Yet that is what the British government should do to prevent emissions remaining excessively high and to fulfil its commitments under the Paris Agreement of 2015.

Next, governments should align all their other policies to conformity with their nationally determined contributions (rather than opening new coal mines or oilfields), and should rachet up these commitments, as was expected after the Paris Agreement but not significantly brought about by the time of COP27 (November 2022). Currently, our planet is liable to an emissions increase corresponding to a temperature rise of well over 2°C, but it will encounter disasters if the likely temperature rise is not limited to 1.5°C (see chapter 2). These commitments should include the mitigation of domestic emissions, reductions of the volume of imported goods generated through emissions elsewhere (such as in China), and domestic adaptation to threats resulting from climate change, such as the risk of floods and of coastal erosion.

Further, the governments of developed countries and of the richer developing countries, such as China and Malaysia, should contribute generously to the new Loss and Damage Fund, agreed at COP27. This would make it possible for the other developing countries to make provision to withstand droughts and floods, and generally to enhance their infrastructures, currently at risk from the impacts of climate change. These changes require funding from resources such as the Loss and Damage Fund and bilateral aid from developed countries, both of a financial and a technological kind. Technology transfer can play a large part here, with many solutions being small-scale or medium-scale ones. In particular, the introduction of large dams (prone to silt up and to undermine local fisheries) should be replaced by sets of small dams, placed so as to prevent floods, facilitate fisheries and generate electricity. Contributions to the Loss and Damage Fund are justified on the basis of obligations to compensate countries that have suffered and are suffering as a result of industrialization and greenhouse gas emissions in developed countries. But if electorates are reluctant to vote for compensation, other terms, such as 'overseas aid' can be used, as long as payments are made in a timely manner and are of sufficient amplitude.

Responsibilities of governments also include matters discussed in earlier sections of this chapter. For example, governments that have not already done so should exit the European Energy Charter, for the reasons already given. They should also play their part in the adoption of ecocide as a crime at the International Criminal Court, alongside the crimes of genocide, crimes against humanity, war crimes and aggression, so that the practices of companies and of relevant individuals are remoulded to avoid related charges. And they should jointly consider whether they can make progress towards the suggestion by E. O. Wilson that half of the Earth's surface should be designated a human-free natural reserve to preserve biodiversity, even if the full implementation of this proposal is impractical (Wilson 2016).

There again, all governments have responsibilities to their own peoples both to mitigate their emissions and to adapt to climate change that cannot be reversed. The ingenious measures adopted by Bangladesh over recent years have been noted earlier in this chapter, as should the innovatory plans of Tuvalu be, to retain some territory of its own in face of the prospect of total inundation (Fainu 2023a, 2023b). Only if virtually all countries participate in emissions reductions can the ceiling of a 1.5°C increase in average temperatures be achieved. Some countries currently engaged in oppressing their own populations, such as Myanmar, Syria and Iran, need first to abandon these practices; only if they do can their peoples be expected to participate in mitigation practices. Yet the wealthier countries can and should assist with such programmes, and their assistance will make it far easier for the governments of developing countries to sell schemes of mitigation and adaptation to their electorates.

Responsibilities of individuals and households

The final paragraphs of chapters 4 and 5 have raised this issue already. Individuals and households should reduce their consumption of meat, and their travel by car and by plane. Despite the limits of offsetting, those who fly should offset their carbon footprint, preferably through charities seeking to preserve or reafforest vulnerable forest systems such as those of Borneo or Canada or the Amazon. Where possible, double glazing and cavity-wall insulation should be adopted and solar panels installed, together with batteries to store the electricity generated, and gas boilers should be replaced or supplemented with heat pumps. (Experience shows that air-based heat pumps are a particularly cheap and efficient way to heat water.) Opportunities provided by local authorities to recycle should be put to good use, as should opportunities for composting.

The policies of local authorities make a large difference to what households can achieve. Many collect recyclable paper, cardboard and recyclable plastic, and (separately) garden waste, and should make provision for collecting these items separately from general rubbish. They should also require adequate insulation in new housing, and retrofit older housing stock to increase energy efficiency. There again, the unnecessary construction of new roads, bypasses and motorways should be avoided; additional roads tend simply to become congested with additional traffic.

But policies such as these, and parallel policies at national level, can only be initiated if there is sufficient public support. Participation in pressure groups can make a crucial difference, as can campaigning within political parties and on the part of political parties for green solutions and policies. Voters cannot consistently complain about the shortcomings of local or national politicians if they prioritize voting for minimal local and national taxes, and this results in cost-cutting policies, rather than socially and environmentally constructive measures.

There is also a role for those who recognize the responsibilities of governments (including devolved governments) and of local authorities to witness to their convictions. Demonstrations and protests (as well as letters to elected representatives) are widely acknowledged to be interpreted as expressing views more widely held than those of the numbers participating. While not all are able to take part in public forms of witness, those able to do so should not lightly pass up opportunities to stand (or sit) and be counted.

Recommended reading

Attfield, Robin. 2015 [1999]. *The Ethics of the Global Environment*. Edinburgh: Edinburgh University Press.

Gardiner, Stephen M. 2011. 'Some Early Ethics of Geoengineering the Climate: A Commentary on the Values of the Royal Society Report'. *Environmental Values* 20(2), 163–88.

Lazarus, Oliver, McDermid, Sonali and Jacquet, Jennifer. 2021. 'The Climate Responsibilities of Industrial Meat and Dairy Producers'. *Climatic Change* 165(1): 30.

McKinnon, Catriona. 2022, *Climate Change and Political Theory*. Cambridge and Hoboken, NJ: Polity Press.

Nolt, John. 2015. *Environmental Ethics for the Long Term: An Introduction*. Abingdon and New York: Routledge.

Oreskes, Naomi and Conway, Eric M. 2010. 'Defeating the Merchants of Doubt'. *Nature* 465 (9 June): 686–7.

Pagliarini, André. 2023. 'Can Lula Save the Amazon? His Record Shows He Might Just Pull it Off'. *Guardian*, 3 January.

Roser, Dominic and Seidel, Christian. 2017. *Climate Justice: An Introduction*. Abingdon and New York: Routledge.

Roupé, Jonas and Ragnarsdøttir, Kristín Vala. 2022. 'Executive Summary', in *Ecocide Law for the Paris Agreement*. Schumacher Foundation, 3.

Wilson, E. O. 2016. *Half-Earth: Our Planet's Fight for Life*. New York: Liveright.

Further reading

Clark, Lesley, Storrow, Benjamin and Harvey, Chelsea. 2023. 'Exxon's Own Models Predicted Global Warming – It Ignored Them', *Scientific American*, 13 January. https://www .scientificamerican.com/article/exxons-own-models-predicted -global-warming-it-ignored-them/

Dias, Peri. 2022. 'Argentinian "Carbon Bomb" Jeopardizes Climate Efforts'. Press release, 350.org, 3 November. https:// 350.org/press-release/vaca-muerta-is-a-carbon-bomb-that -could-eat-up-more-than-11-of-the-global-CO_2-budget/

European Energy Charter, The. 2015. https://www.energycharter .org/process/european-energy-charter-1991/

Fainu, Kalolaine. 2023. 'As sea level rises Tuvalu seeks to tackle the threat of extinction'. *Guardian*, 28 June, 20–2.

Greenfield, Patrick and Weston, Phoebe. 2022. 'Leaders hail CoP15 biodiversity accord – as crisis talks begin to prevent it unravelling'. *Guardian*, 20 December, 11.

Oreskes, Naomi, Conway, Eric M., Karoly, David J., Gergis, Joelle, Neu, Urs and Pfister, Christian. 2018. 'The Denial of Global Warming', in S. White, C. Pfister and F. Mauelshagen (eds), *The Palgrave Handbook of Climate History*. London: Palgrave Macmillan, 149–71.

Ott, Konrad. 2011. 'Domains of Climate Ethics'. *Jahrbuch für Wissenschaft und Ethik* 16: 95–112.

Routley (later Sylvan), Richard. 1973. 'Is There a Need for a New, an Environmental, Ethic?'. *Proceedings of the XVth World Congress of Philosophy*. Varna (Bulgaria): Sofia Press, 205–10. [Reprinted in Robin Attfield (ed.), *The Ethics of the Environment*. Farnham: Ashgate, 2008, 3–12.]

Sierra Club. 2021. 'FAQ: Climate Peace Clause'. https:// tradejusticeedfund.org/wp-content/uploads/FAQ-Climate -Peace-Clause-Final.-1-1.pdf

World Nuclear Association. 2022. 'Chernobyl Accident 1986'. https://world-nuclear.org/information-library/safety-and -security/safety-of-plants/chernobyl-accident.aspx

7

Responding to the Crises

Emotions can block or boost crisis resolution

Crises are severe problems, which may relent or may intensify. Action to confront them can be urgently needed, and I have already argued that this is the case with the climate crisis, the biodiversity crisis and the air pollution crisis. When a crisis requires urgent action, it can also be regarded as an emergency. The three crises discussed here comprise nothing less than a triple emergency.

Yet responses depend to a large extent on the state of mind of ourselves, the respondents and other agents, government ministers included. When risks are severe, a natural reaction is fear, although another is denial. Chapters 2 and 3 were partly written to attest to the genuineness of the crises of climate, biodiversity loss and air pollution, because without a clear grasp of the problems people can fail to respond appropriately. Climate change denial has been a widespread response, albeit perhaps one that is waning, and one needing to be resisted. I have argued in chapters 4, 5 and 6, by contrast, that governments, companies, households and individuals have obligations to mitigate their greenhouse gas emissions and their ecological footprint, and to take concerted action to confront and overcome the crises.

There is, however, a danger that exposing the direness of the situation can generate paralysing fear, together with either inaction or escapism. This possibility has recently been discussed by Jerome Ballet, Damien Bazin and Emmanuel Petit (2023). These authors raise the possibility that fear of ecological collapse may inhibit action and spread pessimism and despair, citing research in different disciplines about the various responses that can be motivated by fear (Moser 2007; Nicholson 2002; Swim et al. 2011).

These same authors discuss the role of emotions such as fear, citing the approach of the pragmatist philosopher John Dewey. In Dewey's later work, emotion guides experience and contributes to the process of inquiry into problematic situations, as well as to responses to those situations. Thus we should not theorize about fear in general, because fear is always 'a fear', focused on a particular situation; and fears also vary between fears that are insufficient (verging on apathy), fears that are excessive (inhibiting action) and fears that are moderate and (in combination with reason) prone to foster proportionate action (Dewey 1934). This last kind of fear can, in situations of danger, actually be desirable. It can also, unlike apathy and unlike terror, be compatible with a qualified form of hope.

Thus Ballet and his fellow authors find helpful Dewey's treatment of emotions such as fear, particularly in view of his rejection of the inevitability of disaster (Dewey 1935: 44). For the sake of rejecting extreme fear, they too reject 'the thesis of collapse', effectively on pragmatist grounds (Ballet, Bazin and Petit 2023: 9). The editor, Nicholas Bardsley, expresses understandable surprise at their grounds for rejection of collapse theory (Bardsley 2023: 1). Yet the version of collapse theory that they reject is that of Y. Cochet, who holds that 'the collapse of the globalized society is . . . certain around 2030, give or take a year' (Ballet, Bazin and Petit 2023: 8: Cochet 2019: 40); and this particular claim can reasonably be rejected on the strength of the available evidence (as presented in chapters 2 and 3 above), indirectly threatening as it remains for later decades of the current century. Meanwhile, the appraisal of ecological fear on the part of Ballet and his fellow authors remains a welcome one.

As they also remark, emotions are capable of preventing 'values' (that is, agents' moral beliefs) from exercising a motivating force. However, as they add, emotions can instead reinforce the motivational force of values, 'giving individuals a greater sense of responsibility' (Ballet, Bazin and Petit 2023: 6). Their salutary critique of the emotion of fear helps to show how fear that is neither insufficient nor excessive can, in combination with ethical principles and with well-judged hope, contribute to solutions of our ecological crises. But so too can love, whether of contemporary human beings, of our descendants, or of the natural world. And likewise so can a sense of solidarity with all people and all creatures.

Deselecting the model of the 'tragedy of the commons'

Yet theories of inevitability can adopt a subtler form. In the same journal number, Jakob Ortmann and Walter Veit appraise the appeal of the widespread framing of the climate crisis and efforts to mitigate it as a 'tragedy of the commons', not least on the part of the Intergovernmental Panel on Climate Change itself (IPCC 2014: 211). This scenario was devised long ago by Garrett Hardin, who pictured herdsmen sharing a pasture of common land. Each is compelled by rational self-interest to increase his herd of cattle, even though this foreseeably leads to the overgrazing of the pasture. Hardin concluded that systems of free access to common resources inevitably bring ruin to all (Hardin 1968: 1244). Yet the framing of climate mitigation as a 'tragedy of the commons' (Ortmann and Veit 2023: 66–8) could itself prove ruinous.

For, as Ortmann and Veit comment, this model, which may seem to explain the failure of humanity to mitigate its greenhouse gas emissions sufficiently, also exhibits 'performativity' (an extension of J. L. Austin's coinage in *Philosophical Papers* [Austin 1962], which originally related to what utterances like 'I promise' thereby perform). In other words, such models are efficacious and self-fulfilling, and liable to foster an attitude of despair, grounded in the imputed inevitability of failure. And this 'performativity' raises the issue of what attitude should be adopted to such models.

Writing in 2011, M. Kopec emphasized that the 'tragedy of the commons' model fails to fit the facts of the situation and should be rejected as false, together with its implications of hopelessness (Kopec 2011: 1249–59; 2023: 203–21). This is in fact a reasonable conclusion, not least because (among other reasons) there have been successful and long-lasting systems of common pasturage and of shared communal resources both in Switzerland and over much of the Third World, sustained by shared norms of self-restraint (as opposed to Hardin's supposed drives of self-interest) (Dietz, Ostrom and Stern 2003; Ostrom 2015; Ostrom et al. 1999). Another ground consists in the observable effectiveness of international efforts to reduce the rate of increase of carbon emissions through the introduction of renewable energy generation. But, as Ortmann and Veit respond to Kopec, all models are literally false in any case and cannot be expected to fit all the facts, and so the falsity of this model is not a conclusive ground against using it for explanatory purposes.

However, as they proceed to argue, we have a choice in the matter of how to frame climate change, and which model to adopt; for all models are also, as W. V. Quine explained, underdetermined by the phenomena (Quine 1951). Besides, there are several alternative models that could be entertained, not all locating rationality in expanding one's country's greenhouse gas emissions. Some of the alternatives, as Kopec had earlier remarked, are

ones in which cooperation is central to the management of
shared resources (Kopec 2011: 216–17). Yet international
negotiations may well have been bedevilled by belief on
the part of negotiators that the model of the tragedy of
the commons applies, that all the participants are driven
by self-interest, and thus that no effective agreement on
mitigation is possible.

Ortmann and Veit accordingly come up with arguments
suggesting that other models are preferable, and that the
assumptions of the 'tragedy of the commons' model are
significantly misguided. Thus the assumption that greenhouse gas emissions are central to the well-being of one's
population, and crucial to the flourishing of national
economies, is highly vulnerable. So continually increasing
these emissions may not be rational. There again, it may
well be rational to 'create binding agreements in order
to limit emissions' (Ortmann and Veit 2023: 78). The
clear advantages for everyone of multilateral mitigation
suggest that some other, more cooperative model should
be adopted, which does not imply that these advantages are unobtainable. There again, the failure of many
attempts to achieve reductions of greenhouse gas emissions
can be explained without resort to the 'tragedy of the
commons' model; for example, the lobbying of petrochemical corporations and other vested interests may well
form part of the explanation (Monbiot 2007: 20–42). The
tragedy model is not just trivially false, as all models are;
it is also demonstrably false. Accordingly, as Ortmann and
Veit argue, it should be consigned to oblivion.

So, too, it may be added, should the model that
frames ecological problems as due to unrelenting human
population growth, and therefore beyond remedy. While
such growth has exacerbated some of the problems, it has
played nowhere near the role that technology and industrial expansion have done. (Compare the pollution levels
of such heavily populated countries as the Netherlands
with those of such more lightly populated areas like the tar
sands regions of Canada.) Besides, the widespread claims

that human population growth is exponential (or geometrical) are simply wrong, for the rate of increase is itself decreasing (making the graph of population across time resemble an 'S' curve). A much saner account of population trends can be found in Hans and Ola Rosling's *Factfulness: Ten Reasons Why We're Wrong About the World – and Why Things Are Better Than You Think* (2018).

Overcoming the obstacles

The partial success of COP26 (2021) and COP27 (2022) in matters of climate change and of COP15 (2022) in matters of biodiversity preservation (despite their shortcomings) suggests that the model of the tragedy of the commons has already begun to slip and to be discarded, and that international negotiators are heeding their responsibilities or the demands of their electorates, or both. Yet new obstacles to constructive agreements and to their implementation have also emerged, such as the war between Russia and Ukraine.

Another obstacle is the brevity of the time frame that many politicians tend to envisage, with the next election variously four years or five years away. Politicians are going to need to take into account future generations, and thus the coming decades and centuries, if global problems are to be confronted. In Wales, this has begun to come about, with the enactment of the Well-Being of Future Generations Act (Welsh Government 2015), and the connected appointment of a Future Generations Commissioner. All this has informed the recently announced decision of the Welsh government to proceed with only 15 of the 51 projected road-building schemes, with the rest being sent back for appraisal of their carbon footprints (BBC 2023e; Wilks 2023: 38). Further, the United Nations itself is about to hold a Futures Summit in 2023, at which a Declaration for Future Generations will be made (Wilks 2023: 38).

Fortunately (and importantly), there are also international pressure groups and websites in being, such as 350.org and Ekō (previously SumOfUs), which alert their participants to the need to act in concert, not least to address hindrances and obstacles to policies of greenhouse gas mitigation and biodiversity preservation. The participants in such networks act as global citizens, aware of their global citizenship. All readers of this book are global citizens, as Nigel Dower explains with respect to all members of humanity (Dower 2007 [1998]); and recognizing that one holds this status can itself foster commitment to action appropriate to the role of global citizen to confront, rather than evade or ignore, the many obstacles to solutions being realized. There are also regional pressure groups, like the African Climate Alliance, a youth-led grassroots campaigning body operating in and around South Africa, advocating for Afro-centric climate justice (African Climate Alliance 2023).

Yet a sense of international solidarity is not the only source of encouragement. Hope is also generated by the positive steps taken, for example at COP27 (November 2022), such as the Loss and Damage Fund (see chapter 6), and at COP15 (December 2022), such as the plan to secure 30 per cent of continents and of oceans for wildlife, and for related international funding (again, see chapter 6). Further, more tangible grounds for hope can be found in the restoration of wild species, such as the return of the red kite first to Wales and then to England, of the golden eagle to Scotland, and of the osprey to the lakes and skies of the United Kingdom as a whole. Meanwhile, the previously falling numbers of puffins are now growing again on islands off the coasts of Wales and Scotland; and this 'good news' story can be replicated many times over. These restorations (and equally welcome ones reported from other countries) are signs that not even biodiversity loss is irreversible, and that well-devised efforts at the restoration of suitable habitats can bring foretastes of natural recovery, capable of being built on further, as long as other problems, such as greenhouse gas emissions, are tackled at the same time.

Thus, among the motivations for commitment to solving our global problems, there is awareness of the need to act before it is too late (see chapters 2 and 3); but there is also awareness of and commitment to the independent value of the flourishing of both human and non-human life, and the principles that this value upholds (chapters 4 and 5), and of the feasibility of social and political solutions (chapter 6). While fear (of the moderate kind), in combination with glimmers of hope, will motivate some (see the opening of the present chapter), inspiration will continue to arise from love of life on Earth, and from the well-grounded hope that it can be perpetuated in all its diversity.

Recommended reading

Ballet, Jerome, Bazin, Damien and Petit, Emmanuel. 2023. 'The Ecology of Fear and Climate Change: A Pragmatist Point of View'. *Environmental Values* 32(1) (February): 5–24.

Bardsley, Nicholas. 2023. 'Some Fears of the Anthropocene'. *Environmental Values* 32(1) (February): 1–4.

Dower, Nigel. 2007 [1998]. *World Ethics: The New Agenda*, 2nd edn. Edinburgh: Edinburgh University Press.

Kopec, M. 2023. 'Game-theory and the Self-fulfilling Climate Tragedy'. *Environmental Values* 26(2): 203–21.

Monbiot, George. 2007. *Heat: How to Stop the Planet Burning*. London: Penguin Books.

Nicholson, S. W. 2002. *The Love of Nature and the End of the World*. Cambridge, MA: MIT Press.

Ortmann, Jakob and Veit, Walter. 2023. 'Theory Roulette: Choosing that Climate Change is not a Tragedy of the Commons'. *Environmental Values* 32(1) (February): 65–89.

Ostrom, E. Burger, J., Field, C. B., Norgaard, R. B. and Policansky, David. 1999. 'Revisiting the Commons: Local Lessons, Global Challenges'. *Science* 284(5412): 278–82.

Rosling, Hans and Rosling, Ola. 2018. *Factfulness: Ten Reasons Why We're Wrong about the World – and Why Things Are Better Than You Think*. London: Hodder and Stoughton.

Swim, N., Stern, P. C., Doherty, D. J., et al. 2011. 'Psychology's

Contribution to Understanding and Addressing Global Climate Change'. *American Psychologist* 66(4): 241–50.

Further reading

Cochet, Y. 2019. *Devant l'effondrement. Essai de collapsologie.* Paris: Les liens qui libèrent.

Dewey, J. 1934. *Art as Experience.* New York: Minton, Balch & Company.

Dewey, J. 1935. *Liberalism and Social Action.* New York: G. P. Putnam.

Dietz, T., Ostrom, E. and Stern, P. 2003, 'The Struggle to Govern the Commons'. *Science* 302(5652): 1907–12.

Hardin, Garrett. 1968. 'The Tragedy of the Commons'. *Science* 162(3859): 1243–8.

Intergovernmental Panel on Climate Change (IPCC). 2014. *Climate Change 2014: Mitigation of Climate Change: Contribution of Working Group III to the Fifth Assessment Report of the Intergovernmental Panel on Climate Change* (ed. O. Edenhofer et al.). Cambridge: Cambridge University Press. https://www.ipcc.ch/site/assets/uploads/2018/02/ipcc_wg3_ar5_full.pdf

Moser, S. C. 2007. 'More Bad News: The Risk of Neglecting Emotional Responses to Climate Change Information', in S. C. Moser and L. Dilling (eds), *Creating a Climate for Change.* New York: Cambridge University Press, 64–80.

Ostrom, E. 2015. *Governing the Commons.* Cambridge: Cambridge University Press.

Welsh Government. 2015. *The Well-Being of Future Generations Act.* https://www.gov.wales/well-being-of-future-generations-wales

Wilks, Rebecca. 2023. 'What Would Your Grandchild Say?' *New Internationalist* 50(2) (March–April): 36–9.

References

African Climate Alliance. 2023. https://africanclimatealliance
.org

Almond, R. E. A., Grooten, M. and Petersen, T. (eds). 2020.
Living Planet Report 2020 – Bending the curve of biodiversity loss. Gland, Switzerland: WWF. https://www.wwf.org
.uk/sites/default/files/2020-09/LPR20_Full_report.pdf

Aristotle. 2000. *Nicomachean Ethics*, trans. and ed. Roger Crisp.
Cambridge and New York: Cambridge University Press.

Arnall, Alex, Hilson, Christopher and McKinnon, Catriona.
2019. 'Climate Displacement and Resettlement: The
Importance of Claims-Making "From Below"'. *Climate
Policy* 19(6): 665–71.

Attfield, Robin. 1995. 'Preferences, Health, Interests and
Value'. *Justifying Value in Nature: Special Topic Issue of the
Electronic Journal of Analytic Philosophy* 3(2) (May): 7–15.

Attfield, Robin. 2005. 'Future Generations: Considering All
the Affected Parties', trans. into Spanish by Adrián Pradier
and Carmen Velayos Castelo as 'Generaciones futuras:
considerando todas las partes afectadas'. *Isegoría* (June):
35–46.

Attfield, Robin. 2014 [2003]. *Environmental Ethics: An
Overview for the Twenty-First Century*, 2nd edn. Cambridge:
Polity.

Attfield, Robin. 2015 [1999]. *The Ethics of the Global
Environment*. Edinburgh: Edinburgh University Press.

Attfield, Robin. 2019 [1995]. *Value, Obligation and Meta-ethics*.
Leiden, NL: Brill.

Attfield, Robin. 2020 [1987]. *A Theory of Value and Obligation*. Abingdon and New York: Routledge.

Attfield, Robin. 2022a. 'Preferences, Interests, Values and Public Health', in T. N. Siti Pariyani et al. (eds), *Bioetika Multidisiplin (Multidisciplinary Perspectives of Bioethics)*. Jakarta: UNHAN RI Press, 189–207.

Attfield, Robin. 2022b. *Applied Ethics: An Introduction*. Cambridge: Polity.

Attfield, Robin and Clutterbuck, John. 2014. 'Climate Refugees, Disappearing States and Territorial Compensation', in L. Chenyang (ed.), *Proceedings of Philosophy: Yesterday, Today and Tomorrow Conference*. Singapore: Philosophy Yesterday, Today and Tomorrow Conference.

Attfield, Robin and Humphreys, Rebekah. 2016. 'Justice and Non-human Animals, Part I'. *Bangladesh Journal of Bioethics* 7(3): 1–11.

Attfield, Robin and Humphreys, Rebekah. 2017. 'Justice and Non-human Animals, Part II'. *Bangladesh Journal of Bioethics* 8(1): 44–57.

Austin, J. L. 1962. *How to Do Things with Words*. Oxford: Clarendon Press.

Baer, Paul, Athanasiou, Tom, Kartha, Sivan and Kemp-Benedict, Eric 2008. *The Greenhouse Development Rights Framework: The Right to Development in a Climate-Constrained World*, 2nd edn. Berlin: Heinrich Böll Foundation, Christian Aid, Eco-Equity and the Stockholm Environment Institute.

Ballet, Jerome, Bazin, Damien and Petit, Emmanuel. 2023. 'The Ecology of Fear and Climate Change: A Pragmatist Point of View'. *Environmental Values* 32(1) (February): 5–24.

Bardsley, Nicholas. 2023. 'Some Fears of the Anthropocene'. *Environmental Values* 32(1) (February): 1–4.

Barnosky, Anthony D. 2014. *Dodging Extinction: Power, Food, Money and the Future of Life on Earth*. Oakland, CA: University of California Press.

Barnosky, Anthony D. 2017 [2015]. 'Five Climate Tipping Points We've Already Seen, and One We're Hoping For'. Blog. https://www.huffpost.com/entry/five-climate-tipping-poin_b_8166588

BBC. 2019. 'Ella Kissi-Debrah: New Inquest into Girl's "Pollution" Death', 2 May. https://www.bbc.co.uk/news/uk-england-london-48132490

BBC. 2020. 'Ella Adoo-Kissi-Debrah: Air Pollution a Factor in Girl's Death, Inquest Finds', 16 December. https://www.bbc.co.uk/news/uk-england-london-55330945

BBC. 2021. 'COP26: What Was Agreed at the Glasgow Climate Conference?' BBC News: Science & Environment, 15 November. https://www.bbc.co.uk/news/science-environment-56901261

BBC. 2022a. Radio 4 and BBC1 News, 17 and 18 July.

BBC. 2022b. Radio 4 News, 5 August.

BBC. 2022c. 'Newsnight' (BBC2), late August.

BBC. 2023a. Radio 4 News. High Seas Treaty, 6 March.

BBC. 2023b. *World at One*, 31 January.

BBC. 2023c. Radio 4 News. Chris Skidmore MP's report on decarbonizing the economy, 13 January.

BBC. 2023d. Radio 4. Justin Rowlatt in broadcast on steel production at Port Talbot, 23 January.

BBC. 2023e. Radio 4 News. Welsh government road-building schemes, 14 February.

Bednar-Friedl, Birgit, Biesbroek, Robert and Schmidt, Daniela N. 2022. 'Europe', in H.-O. Pörtner et al. (eds), *Climate Change 2022: Impacts, Adaptation and Vulnerability*. Contribution of Working Group II to the Sixth Assessment Report of the Intergovernmental Panel on Climate Change. Cambridge and New York: Cambridge University Press, 1817–1927. https://www.ipcc.ch/report/ar6/wg2/downloads/report/IPCC_AR6_WGII_Chapter13.pdf

Belcher, Stephen. 2022. Presentation to House of Commons, Westminster, 12 July.

Biermann, Frank and Boas, Ingrid. 2008. 'Protecting Climate Refugees: The Case for a Global Protocol'. *Environment: Science and Policy for Sustainable Development* 50(6): 8–17.

Bittel, Jason. 2022. 'Saving Big Mammals Fights Extinctions and Climate Change'. Natural Resources Defense Council, 23 June. https://www.nrdc.org/stories/saving-big-mammals-fights-extinction-and-climate-change?source=EMOCT15INF&utm_source=alert&utm_medium=text&utm_campaign=email

Blaylock, Jean. 2022. 'What is the Energy Charter Treaty and why do we need to exit?' *Global Justice Now*, 24 June. https://www.globaljustice.org.uk/blog/2022/06/energy-charter-treaty/

Breaking Boundaries: The Science of Our Planet [film]. 2021. David Attenborough and Johan Rockström, dir. Jon Clay. All3Media.

Broome, John. 2012. *Climate Matters: Ethics in a Warming World*. New York and London: W. W. Norton & Co.

Brown, David. 2019. 'Climate Justice and REDD+: A Multiscalar Examination of the Norwegian–Ethiopian Partnership', in Tahseen Jafry (ed.), *Routledge Handbook of Climate Justice*. Abingdon and New York: Routledge, 262–75.

Brown, Donald A. et al. 2005. *White Paper on the Ethical Dimensions of Climate Change*. Rock Ethics Institute, Pennsylvania State University.

Brown, Kathryn. 2023. Presentation on COP27 and COP15 to the Cardiff Branch of The Wildlife Trusts, 19 January.

Caldecott, Julian. 2022a. 'Implications of Earth System Tipping Pathways for Climate Change Mitigation Investment' (working paper, June). Bristol: Schumacher Institute for Sustainable Systems.

Caldecott, Julian. 2022b. 'Implications of Earth System Tipping Pathways for Climate Change Mitigation Investment'. *Discover Sustainability* 3 (8 November): https://link.springer.com/article/10.1007/s43621-022-00105-7

Caney, Simon. 2010. 'Climate Change, Human Rights and Moral Thresholds', in Stephen M. Gardiner, Simon Caney, Dale Jamieson and Henry Shue (eds), *Climate Ethics*. Oxford and New York: Oxford University Press, 163–77.

Caney, Simon. 2012. 'Just Emissions'. *Philosophy & Public Affairs* 40(4): 255–300.

Caney, Simon. 2020. 'The Right to Resist Global Injustice', in Thom Brooks (ed.), *The Oxford Handbook of Global Justice*. Oxford: Oxford University Press, 510–36.

Carrington, Damian. 2022. 'World is Coming Close to Irreversible Change, Say Climate Experts'. *Guardian*, 28 October, 4.

Carson, Rachel. 2000 [1962]. *Silent Spring*. London: Penguin Classics.

Center for Biological Diversity. 2022a. 'Center Op-Ed: Extreme Heat is Driving Extinction'. *Endangered Earth* 1151, 28 July. https://www.biologicaldiversity.org/publications/earthonline/endangered-earth-online-no1151.html

Center for Biological Diversity. 2022b, 'IUCN: Monarchs

Are Endangered'. https://www.biologicaldiversity.org /publications/earthonline/endangered-earth-online-no1151.html

Cho, Renée. 2021. 'How Close Are We to Climate Tipping Points?' *State of the Planet*, 11 November. Columbia Climate School. https://news.climate.columbia.edu/2021/11/11/how -close-are-we-to-climate-tipping-points

Clark, Lesley, Storrow, Benjamin and Harvey, Chelsea. 2023. 'Exxon's Own Models Predicted Global Warming – It Ignored Them'. *Scientific American*, 13 January. https://www .scientificamerican.com/article/exxons-own-models-predicted -global-warming-it-ignored-them/

Climate Variability Forum. 2010. *Climate Variability Monitor 2010: The State of the Climate Crisis*. New York: DARA.

Cochet, Y. 2019. *Devant l'effondrement. Essai de collapsologie*. Paris: Les liens qui libèrent.

Cripps, Elizabeth. 2022. *What Climate Justice Means and Why We Should Care*. London: Bloomsbury Continuum.

Crocker, David A. 1996, 'Hunger, Capability and Development', in William Aiken and Hugh LaFollette (eds), *World Hunger and Morality*, 2nd edn. Upper Saddle River, NJ: Prentice-Hall, 211–30.

Data for Progress. 2022. *DFP Newsletter*, email communication received 9 July.

Defra (UK). 2022. 'UK Air: Effects of Air Pollution'. https:// uk-air.defra.gov.uk/air-pollution/effects

Dewey, J. 1934. *Art as Experience*. New York: Minton, Balch & Company.

Dewey, J. 1935. *Liberalism and Social Action*. New York: G. P. Putnam.

Dias, Peri. 2022. 'Argentinian "Carbon Bomb" Jeopardizes Climate Efforts'. Press release, 350.org, 3 November. https:// 350.org/press-release/vaca-muerta-is-a-carbon-bomb-that -could-eat-up-more-than-11-of-the-global-co₂-budget/

Dietz, T., Ostrom E. and Stern P. 2003. 'The Struggle to Govern the Commons'. *Science* 302(5652): 1907–12.

Dower, Nigel. 2007 [1998]. *World Ethics: The New Agenda*, 2nd edn. Edinburgh: Edinburgh University Press.

Dworkin, Ronald. 1977. *Taking Rights Seriously*. London: Duckworth.

El Hinnawi, Essam. 1985. *Environmental Refugees*. Nairobi: United Nations Environment Programme.

Encyclopaedia Britannica. 2020. 'K-T Extinction'. *Encyclopaedia Britannica*. https://www.britannica.com/science/K-T-extinction

Ericson, Jason P., Vörösmarty, Charles J., Dingman, S. Lawrence, Ward, Larry G. and Meybeck, Michel. 2006. 'Effective Sea-level Rise and Deltas: Cause of Change and Human Dimension Implications'. *Global and Planetary Change* 50: 63–82.

European Energy Charter. 2015. https://www.energycharter.org/process/european-energy-charter-1991/

Fainu, Kalolaine. 2023a. 'Facing Extinction, Tuvalu Considers the Digital Clone of a Country', 27 June. https://www.theguardian.com/world/2023/jun/27/tuvalu-climate-crisis-rising-sea-levels-pacific-island-nation-country-digital-clone

Fainu, Kalolaine. 2023b. 'As Sea Level Rises Tuvalu Seeks to Tackle the Threat of Extinction'. *Guardian*, 28 June, 20–1.

Francis, Blake. 2020. 'Climate Change Injustice'. *Environmental Ethics* 44(1) (Spring): 5–24.

Friends of the Earth. 2022. 'Monarch Butterflies Are Starving'. February: https://www.similarmail.com/edm-newsletter/foe.org/5998018_your-signature-is-needed-monarch-butterflies-are-starving.html

Fuller, Gary. 2022. 'Even Low Levels of Air Pollution Can Damage Health, Study Finds'. *Guardian*, 12 August. https://www.theguardian.com/environment/2022/aug/12/even-low-levels-of-air-pollution-can-damage-health-study-finds

Gardiner, Stephen M. 2011. 'Some Early Ethics of Geoengineering the Climate: A Commentary on the Values of the Royal Society Report'. *Environmental Values* 20(2): 163–88.

Goodpaster, Kenneth E. 1978. 'On Being Morally Considerable'. *Journal of Philosophy* 75: 308–25.

Goodpaster, Kenneth E. 1980. 'On Stopping at Everything: A Reply to W. M. Hunt'. *Environmental Ethics* 2(3): 281–4.

Graham, Hilary, Bland, J. Martin, Cookson, Richard, Kanaan, Mona and White, Piran C. L. 2017. 'Do People Favour Policies that Protect Future Generations? Evidence from a British Survey of Adults'. *Journal of Social Policy* 46(3) (July): 23–445.

Gramling, Carolyn. 2021. 'Australian Fires in 2019–2020 had Even More Global Reach than Previously Thought'. *Science News*, 15 September. Washington, DC: Society for Science.

https://www.sciencenews.org/article/australia-wildfires
-climate-change-carbon-dioxide-ocean-algae

Granberg-Michaelson, Wesley. 1992. *Redeeming the Creation: The Rio Earth Summit – Challenges for the Churches*. Geneva: WCC Publications.

Greenfield, Patrick and Weston, Phoebe. 2022a. 'Cop 15: Historic Deal Struck to Halt Biodiversity Loss by 2030', *Guardian*, 19 December. https://www.theguardian.com/environment/2022/dec/19/cop15-historic-deal-signed-to-halt-biodiversity-loss-by-2030-aoe

Greenfield, Patrick and Weston, Phoebe. 2022b. 'Leaders Hail CoP15 Biodiversity Accord – as Crisis Talks Begin to Prevent It Unravelling'. *Guardian*, 20 December, 11.

Guardian. 2020. 'Kiribati's President's Plans to Raise Islands in Fight Against Sea-Level Rise'. *Guardian*, 10 August. https://www.theguardian.com/world/2020/aug/10/kiribatis-presidents-plans-to-raise-islands-in-fight-against-sea-level-rise

Guterres, António. 2021. 'UN Secretary-General's Foreword', in *Making Peace with Nature: A Scientific Blueprint to Tackle the Climate, Biodiversity and Pollution Emergencies*. Nairobi: United Nations Environmental Programme.

Hammond, Allen L. (ed.). 1994. *World Resources, 1994–5*. Oxford and New York: Oxford University Press.

Hardin, Garrett. 1968. 'The Tragedy of the Commons'. *Science* 162(3859): 1243–8.

Hare, R. M. 1981. *Moral Thinking: Its Levels, Methods and Point*. Oxford: Clarendon Press.

Hare, R. M. 1996. 'Why I am Only a Demi-Vegetarian', in R. M. Hare, *Essays on Bioethics*. Oxford: Oxford University Press, 219–35.

Harvey, Fiona. 2022a. 'Major Cities Blighted by Nitrogen Dioxide Pollution, Research Finds'. *Guardian*, 17 August. https://www.theguardian.com/environment/2022/aug/17/major-cities-blighted-by-nitrogen-dioxide-pollution-research-finds

Harvey, Fiona. 2022b. '"More than 50 Poor Countries in Danger of Bankruptcy" Says UN Official'. *Guardian*, 10 November. https://www.theguardian.com/environment/2022/nov/10/54-poor-countries-in-danger-of-bankruptcy-amid-economic-climate-cop27

Health Effects Institute. 2023. *State of Global Air.* https://www
.stateofglobalair.org

Helmore, Edward. 2022. '"Time Has Run Out": UN Fails to
Reach Agreement to Protect Marine Life'. *Guardian,* 27
August. https://www.theguardian.com/world/2022/aug/27
/united-nations-ocean-treaty-marine-life

Henderson, Gideon. 2022. Presentation to House of Commons,
Westminster, 12 July.

Heyward, Clare and Ödalen, Jörgen. 2016. 'A Free Movement
Passport for the Territorially Dispossessed', in Clare Heyward
and Dominic Roser (eds), *Climate Justice in a Non-Ideal
World.* Oxford: Oxford University Press, 208–26.

Horton, Helena. 2022. 'Atmospheric Levels of All Three
Greenhouse Gases Hit Record High'. *Guardian,* 26 October.
https://www.theguardian.com/environment/2022/oct/26/
atmospheric-levels-greenhouse-gases-record-high

Houghton, John. 2015. *Global Warming: The Complete
Briefing,* 5th edn. Cambridge: Cambridge University Press.

Hourdequin, Marion. 2007. 'Doing, Allowing and Precaution'.
Environmental Ethics 29(4): 339–58.

Huq, Saleemul. 2023. Contribution to 'Pledges and Progress',
episode 1 of *Rethink Climate,* BBC Radio 4, 2 January.

IPCC (Intergovernmental Panel on Climate Change). 2013.
'Summary for Policymakers', in T. F. Stocker et al. (eds),
*Climate Change 2013: The Physical Science Basis:
Contribution of Working Group I to the Fifth Assessment
Report of the Intergovernmental Panel on Climate Change.*
Cambridge: Cambridge University Press, 3–29. https://www
.ipcc.ch/site/assets/uploads/2018/02/WG1AR5_all_final.pdf

IPCC (Intergovernmental Panel on Climate Change). 2014.
*Climate Change 2014: Mitigation of Climate Change:
Contribution of Working Group III to the Fifth Assessment
Report of the Intergovernmental Panel on Climate Change*
(ed. O. Edenhofer et al.). Cambridge: Cambridge University
Press. https://www.ipcc.ch/site/assets/uploads/2018/02/ipcc
_wg3_ar5_full.pdf

IPCC (Intergovernmental Panel on Climate Change). 2021.
Summary for Policymakers. In: *Climate Change 2021: The
Physical Science Basis. Contribution of Working Group I to
the Sixth Assessment Report of the Intergovernmental Panel
on Climate Change.* Cambridge University Press: Cambridge

and New York, pp. 3–32. https://www.cambridge.org /core/books/climate-change-2021-the-physical-science-basis /summary-for-policymakers/8E7A4E3AE6C364220F3B76A 189CC4D4C

IPCC (Intergovernmental Panel on Climate Change). 2023. 'Urgent Climate Action Can Secure a Liveable Future for All: AR6 Synthesis Report: Climate Change 2023', 20 March, https://www.ipcc.ch/2023/03/20/press-release-ar6-synthesis -report/

IUCN (International Union for Conservation of Nature). 2022a. *IUCN Red List of Threatened Species.* https://www .iucnredlist.org

IUCN (International Union for Conservation of Nature). 2022b. 'Migratory Monarch Butterfly Now Endangered – IUCN Red List', 21 July. https://www.iucn.org/press-release/202207 /migratory-monarch-butterfly-now-endangered-iucn-red-list

Jafry, T. (ed.). 2019. *Routledge Handbook of Climate Justice.* Abingdon and New York: Routledge.

Kelbessa, Workineh. Forthcoming. 'Climate Justice for Africa'.

Kolbert, Elizabeth. 2014. *The Sixth Extinction: An Unnatural History.* New York: Henry Holt & Co.

Koop, Fermin, Lam, Regina and Zhijian, Xia. 2022. 'COP15 Reaches Historic Agreement to Protect Biodiversity', 21 December. https://chinadialogue.net/en/nature/cop15-reaches -historic-agreement-to-protect-biodiversity/

Kopec, M. 2011. 'A More Fulfilling (and Frustrating) Take on Reflexive Predictions.' *Philosophy of Science* 78: 1249–59.

Kopec, M. 2023. 'Game-theory and the Self-fulfilling Climate Tragedy'. *Environmental Values* 26(2): 203–21.

Kriegler, E., Hall, J. W., Held, H., Dawson, R. and Schellnhuber, H. J. 2009. 'Imprecise Probability Assessment of Tipping Points in the Climate System'. *Proceedings of the National Academy of Science USA* 106: 4133–7.

Lakes Against Nuclear Dump. 2023. Letter to the editor. *Whitehaven News*, January.

Lazarus, Oliver, McDermid, Sonali and Jacquet, Jennifer. 2021. 'The Climate Responsibilities of Industrial Meat and Dairy Producers'. *Climatic Change* 165(1): 30.

Lear, Edward. 2023 [1871]. 'The Owl and the Pussy-Cat', in Poetry Foundation. https://www.poetryfoundation.org/poems /43188/the-owl-and-the-pussy-cat

Lejano, Raul P. and Nero, Shondel. 2020. *The Power of Narrative: Climate Skepticism and the Deconstruction of Science*. New York: Oxford University Press.

Lenton, Timothy M. 2011. 'Early Warning of Climate Tipping Points'. *Nature Climate Change* 1 (July): 201–9.

Lenton, Timothy M., Rockström, Johan, Gaffney, Owen, et al. 2019. 'Climate Tipping Points – Too Risky to Bet Against'. *Nature* 575: 592–5. https://doi.org/10.1038/d41586-019-03595-0

Lie, Livia. 2023. 'Patagonia Needs You to Fight for Climate Justice'. Email communication, 19 January.

Lomborg, Bjørn. 2001. *The Skeptical Environmentalist: Measuring the Real State of the World*. Cambridge: Cambridge University Press.

MacAskill, William. 2022. *What We Owe the Future: A Million-Year View*. London: Oneworld.

Masefield, John. 2005 (1902). 'Sea-Fever', in National Poetry Library 2005. https://www.nationalpoetrylibrary.org.uk/poems/sea-fever

Mathews, Freya. 2010. 'Planetary Collapse Disorder: The Honeybee as Portent of the Limits of the Ethical'. *Environmental Ethics* 32(4): 353–67.

McAllister, Sean. 2022. 'There Could Be 1.2 Billion Climate Refugees by 2050'. Zürich Insurance Group. https://www.zurich.com/en/media/magazine/2022/there-could-be-1-2-billion-climate-refugees-by-2050-here-s-what-you-need-to-know

McKinnon, Catriona. 2022. *Climate Change and Political Theory*. Cambridge and Hoboken, NJ: Polity.

Meinshausen, Malte et al. 2009. 'Greenhouse-Gas Emission Targets for Limiting Global Warming to 2°C'. *Nature* 458 (30 April): 1158–62.

Meyer, Aubrey. 2005. *Contraction & Convergence: The Global Solution to Climate Change*. Totnes: Green Books.

Mims, Christopher. 2009. 'Plan Bee: As Honeybees Die Out, Will Other Species Take Their Place?' *Scientific American* (31 March). http://www.scientificamerican.com/article.cfm?id=other-bee-species-subbing-for-honeybees

Monbiot, George. 2007. *Heat: How to Stop the Planet Burning*. London: Penguin Books.

Mongabay. 2006. *Rainforest Diversity – Origins and*

Implications, 27 July. https://rainforests.mongabay.com
/0301.htm

Moser, S. C. 2007. 'More Bad News: The Risk of Neglecting
Emotional Responses to Climate Change Information', in
S. C. Moser and L. Dilling (eds), *Creating a Climate for
Change*. New York: Cambridge University Press, 64–80.

Mulhern, Owen. 2020. *The Statistics of Biodiversity Loss
[2020 WWF Report]*. https://earth.org/data_visualization
/biodiversity-loss-in-numbers-the-2020-wwf-report/

Myers, Norman. 2005. 'Environmental Refugees: An Emergent
Security Issue'. Prague, 13th Economic Forum (23–7 May).
https://www.osce.org/files/f/documents/c/3/14851.pdf

National Drought Mitigation Center 2022. *US Drought
Monitor: Current Map*. University of Nebraska-Lincoln.
https://droughtmonitor.unl.edu

Newman, J. A., Varner, G. and Linquist S. 2017. *Defending
Biodiversity: Environmental Science and Ethics*. Cambridge:
Cambridge University Press.

Nicholson, S. W. 2002. *The Love of Nature and the End of the
World*. Cambridge, MA: MIT Press.

Nine, Cara. 2010. 'Ecological Refugees, States' Borders and the
Lockean Proviso'. *Journal of Applied Philosophy* 27(4): 359–75.

Nolt, John. 2015. *Environmental Ethics for the Long Term: An
Introduction*. Abingdon and New York: Routledge.

Nordhaus, William. 2008. *A Question of Balance: Weighing the
Options on Global Warming Policies*. New Haven, CT: Yale
University Press.

Norton, Bryan. 1991. *Toward Unity Among Environmentalists*.
New York and Oxford: Oxford University Press.

O'Neill, Onora. 1986. *Faces of Hunger: An Essay on Poverty,
Hunger and Development*. London: Allen & Unwin.

Oreskes, Naomi and Conway, Eric M. 2010. 'Defeating the
Merchants of Doubt'. *Nature* 465 (9 June): 686–7.

Oreskes, Naomi, Conway, Eric M., Karoly, David J., Gergis,
Joelle, Neu, Urs and Pfister, Christian. 2018. 'The Denial of
Global Warming', in S. White, C. Pfister and F. Mauelshagen
(eds), *The Palgrave Handbook of Climate History*. London:
Palgrave Macmillan, 149–71.

Ortmann, Jakob and Veit, Walter. 2023. 'Theory Roulette:
Choosing that Climate Change is not a Tragedy of the
Commons'. *Environmental Values* 32(1) (February): 65–89.

Ostrom, E. 2015. *Governing the Commons*. Cambridge: Cambridge University Press.

Ostrom, E., Burger, J., Field, C. B., Norgaard, R. B. and Policansky, David. 1999. 'Revisiting the Commons: Local Lessons, Global Challenges'. *Science* 284(5412): 278–82.

Ott, Konrad. 2011. 'Domains of Climate Ethics'. *Jahrbuch für Wissenschaft und Ethik* 16: 95–112.

Pagliarini, André. 2023. 'Can Lula Save the Amazon? His Record Shows He Might Just Pull it Off'. *Guardian*, 3 January.

Parfit, Derek. 1984. *Reasons and Persons*. Oxford: Clarendon Press.

Paris Agreement. 2015. Preamble. https://unfccc.int/files /essential_background/convention/application/pdf/english _paris_agreement.pdf

Persson, Erik. 2008. *What is Wrong with Extinction?* Doctoral thesis, Lund: University of Lund.

Pidgeon, Nicholas. 2022. 'A Sustainable Future for Wales?' Unpublished presentation to South Wales Quakers and Cardiff and District United Nations Association, 27 September.

Quine, V. W. 1951. 'Main Trends in Recent Philosophy: Two Dogmas of Empiricism'. *Philosophical Review* 60(1): 20–43.

Ramirez, Rachel. 2021. 'The World's Largest Intact Forest Is in Danger. Here's How to Save It'. *Huffington Post*, 4 March.

Rawls, John. 1972. *A Theory of Justice*. Oxford: Oxford University Press.

Read, Rupert. 2022. *Why Climate Breakdown Matters*. London: Bloomsbury Academic.

Rigaud, Konte Kumari et al. 2018. *Groundswell: Preparing for Internal Climate Migration*. Washington, DC: World Bank.

Rocklöv, Joacim and Dubrow, Robert. 2020. 'Climate Change: An Enduring Challenge for Vector-Borne Disease Prevention and Control'. *Nature Immunology* 21 (20 April): 479–83. https://doi.org/10.1038/s41590-020-0648-y

Roser, Dominic and Seidel, Christian. 2017. *Climate Justice: An Introduction*. Abingdon and New York: Routledge.

Rosling, Hans and Rosling, Ola. 2018. *Factfulness: Ten Reasons Why We're Wrong About the World – and Why Things Are Better Than You Think*. London: Hodder and Stoughton.

Roupé, Jonas and Ragnarsdøttir, Kristín Vala. 2022. 'Executive Summary', in *Ecocide Law for the Paris Agreement*. Schumacher Foundation.

Routley [Sylvan], Richard. 1973. 'Is There a Need for a New, an Environmental, Ethic?' *Proceedings of the World Congress of Philosophy*. Varna (Bulgaria), 205–10. [Reprinted in Robin Attfield (ed.), *The Ethics of the Environment*. Farnham: Ashgate, 2008, 3–12.]

Scherer, Donald. 1983. 'Anthropocentrism, Atomism and Environmental Ethics', in Donald Scherer and Thomas Attig (eds), *Ethics and the Environment*. Englewood Cliffs, NJ: Prentice-Hall, 73–81.

Schramm, Stephen. 2020. 'Blue Devil of the Week: Fighting Extinction with Science'. *Duke Today*, 6 January. https://today.duke.edu/2020/01/blue-devil-week-fighting-extinction-science

Schwartz, Thomas. 1978. 'Obligations to Posterity', in R. I. Sikora and Brian Barry (eds), *Obligations to Future Generations*. Philadelphia, PA: Temple University Press.

Schwartz, Thomas. 1979. 'Welfare Judgements and Future Generations'. *Theory and Decision* 11: 181–94.

Science Daily. 2011. 'How Many Species on Earth? About 8.7 Million, New Estimate Says', 24 August. https://www.sciencedaily.com/releases/2011/08/110823180459.htm

Secretariat of the Antarctic Treaty. 1991. *Protocol on Environmental Protection to the Antarctic Treaty*. https://www.ats.aq

Shue, Henry. 1996. *Basic Rights: Subsistence, Affluence and U.S. Foreign Policy*. Princeton, NJ: Princeton University Press.

Sierra Club. 2021. 'FAQ: Climate Peace Clause'. https://tradejusticeedfund.org/wp-content/uploads/FAQ-Climate-Peace-Clause-Final.-1-1.pdf

Singer, Peter. 1976. *Animal Liberation: A New Ethic for Our Treatment of Animals*. London: Jonathan Cape.

Singer, Peter. 2002. *One World: The Ethics of Globalization*. New Haven, CT and London: Yale University Press.

Singer, Peter. 2023. *Ethics in the Real World: Essays on Things That Matter*. Princeton, NJ and Oxford: Princeton University Press.

Soued, C., Harrison, J. A., Mercier-Blais, S. and Prairie, Yves T. 2022. 'Reservoir CO_2 and CH_4 emissions and their Climate Impact over the Period 1900–2060'. *Nature Geoscience* 15: 700–5.

Stancil, Kenny. 2022. 'Collective Action or Collective Suicide:

UN Chief Pleads for Real Climate Response'. *Common Dreams*, July 18. https://www.commondreams.org/2022/07/18/collective-action-or-collective-suicide-un-chief-pleads-real-climate-response

Stern, Nicholas et al. 2007. *The Economics of Climate Change: The Stern Review*. Cambridge: Cambridge University Press.

Stop Ecocide International. 2022. 'Ecocide Report' (monthly podcast). https://www.stopecocide.earth/the-ecocide-report

Swim, N., Stern, P. C., Doherty, D. J., et al. 2011. 'Psychology's Contribution to Understanding and Addressing Global Climate Change'. *American Psychologist* 66(4): 241–50.

Thomson, Ashley. 2020. 'Biodiversity and the Amazon Rainforest', 22 May. Washington, DC: Greenpeace USA. https://www.greenpeace.org/usa/biodiversity-and-the-amazon-rainforest/

Tollefson, J. 2022. 'Scientists Raise Alarm over "Dangerously Fast" Growth in Atmospheric Methane'. *Nature*, 8 February. https://doi.org/10.1038/d41586-022-00312-2

Traherne, Thomas. 2023. 'Wonder', in Poetry Foundation 2023. https://www.poetryfoundation.org/poems/45418/wonder-56d22507c0b42

UCSD (University of California at San Diego) and SIO (Scripps Institute of Oceanography). 2022. *The Keeling Curve*. UCSD and SIO. https://keelingcurve.ucsd.edu

UNEP (United Nations Environment Programme). 2021. *Why are Coral Reefs Dying?*, 12 November. https://www.unep.org/news-and-stories/story/why-are-coral-reefs-dying

UNESCO. 2017. *Declaration of Ethical Principles in Relation to Climate Change*. Paris: UNESCO.

UNICEF. 2022. 'At Least 10 Million Children Face Severe Drought in the Horn of Africa', 22 April. https://www.unicef.org/press-releases/least-10-million-children-face-severe-drought-horn-africa-unicef

United Nations. 1992. *The Rio Declaration on Environment and Development*. UN Document A/CONF.151/26. New York: United Nations.

United Nations. 2022. 'Intergovernmental Conference on Marine Biodiversity of Areas Beyond National Jurisdiction', 15 August. https://www.un.org/bbnj

Varley, Steve. 2022. 'COP27 – Five Key Takeaways from the UN Climate Talks'. *Economist Impact*, 18 November. http://

impact.economist.com/sustainability/five-key-takeaways
-cop27

Vynne, Carly et al. 2022. 'An Ecoregion-Based Approach to Restoring the World's Intact Large Mammal Assemblages'. *Ecography*: e06098 (Restoration Special Issue). https://onlinelibrary.wiley.com/doi/pdf/10.1111/ecog.06098

Warnock, Geoffrey J. 1971. *The Object of Morality*. London: Methuen.

We Move Europe. 2023. 'A Farming Revolution'. Email message, 15 January.

Welsh Government. 2015. *Well-Being of Future Generations (Wales) Act 2015*. https://www.gov.wales/well-being-of-future-generations-wales

Westra, Laura. 2009. *Environmental Justice and the Rights of Ecological Refugees*. London and Sterling, VA: Earthscan.

Wilks, Rebecca. 2023. 'What Would Your Grandchild Say?' *New Internationalist* 50(2) (March–April): 36–9.

Williams, Mary B. 1978. 'Discounting versus Maximum Sustainable Yield', in R. I. Sikora and Brian Barry (eds), *Obligations to Future Generations*. Philadelphia, PA: Temple University Press, 169–85.

Wilson, E. O. 2016. *Half-Earth: Our Planet's Fight for Life*. New York: Liveright.

Wingspread Statement. 1998. http://www.gdrc.org/u-gov/precaution-3.html

Winters, Joseph. 2022. 'Microplastics in the Ocean: A Climate Change Crisis We Ignore'. *Morning Star*, 14 November, 11.

World Health Organization. 2009. *Global Health Risks: Mortality and Burden of Disease Attributable to Selected Major Risks*. Geneva: World Health Organization. https://apps.who.int/iris/handle/10665/44203

World Health Organization. 2021a. 'What Are the WHO Air Quality Guidelines?' 22 September. https://www.who.int/news-room/feature-stories/detail/what-are-the-who-air-quality-guidelines

World Health Organization. 2021b. 'Ambient (Outdoor) Air Pollution', 22 September. https://www.who.int/news-room/fact-sheets/detail/ambient-(outdoor)-air-quality-and-health

World Nuclear Association. 2022. 'Chernobyl Accident 1986'. https://world-nuclear.org/information-library/safety-and-security/safety-of-plants/chernobyl-accident.aspx

WWF (World Wide Fund for Nature). 2022. 'What Is the Sixth Mass Extinction and What Can We Do About It?' https://www.worldwildlife.org/stories/what-is-the-sixth-mass -extinction-and-what-can-we-do-about-it

Yaron, Lee and Reuters. 2022. 'A 117 Degree Day in Portugal: Record-Breaking Heat Waves Hit Europe'. *Ha-Aretz* (World News, Europe), 18 July.

YPTE (Young People's Trust for the Environment). 2022. *Rainforests: Why Are They Important?* https://ypte.org.uk /factsheets/rainforests/what-are-the-threats-to-the-rainforests

Zinni, Yasmin. 2018. 'Human Influences on the Temperate Rainforest', 25 April. https://sciencing.com/human-influences -temperate-rainforest-8480768.html

Zissu, Alexandra. 2022. 'Colony Collapse Disorder: Why Are Bees Dying?', 29 April. Natural Resources Defense Council. https://www.nrdc.org/stories/buzz-about-colony-collapse -disorder

Index